C000302312

Reality Media

Reality Media

Jay David Bolter, Maria Engberg, and Blair MacIntyre

The MIT Press
Cambridge, Massachusetts
London, England

© 2021 Massachusetts Institute of Technology

All rights reserved. No part of this book may be reproduced in any form by any electronic or mechanical means (including photocopying, recording, or information storage and retrieval) without permission in writing from the publisher.

This book was set in Stone Serif and Stone Sans by Jen Jackowitz. Printed and bound in the United States of America.

Library of Congress Cataloging-in-Publication Data is available.

ISBN: 978-0-262-04512-4

10 9 8 7 6 5 4 3 2 1

For David Bolter
For Fredrik Engberg
For Elizabeth Mynatt

Contents

Preface

This is a book about augmented reality (AR) and virtual reality (VR). We focus on AR and VR not as technologies, but as part of our complex digital media culture. This perspective has determined what to include and what to omit. Our aim is to show how AR and VR function as media and serve as platforms for existing and new forms of expression, rather than to account fully for their varied history and present usage. We will describe how AR and VR relate to and remediate other, more established media, such as film, television, and games.

The book grew out of our own experience with AR and VR as researchers and teachers. In both those roles, we have necessarily been concerned with the characteristics of AR and VR as technologies. But our main concern has been to explore their affordances and overcome their limitations in order to design effective experiences in the areas of education, storytelling, personal expression, and, more recently, communication and conferencing. In the course of this work, we found ourselves explicitly or implicitly appealing to film, video and television, and screen-based video games, and we became convinced of the value of this comparative approach.

Reflecting on how we could elucidate our approach in a book was what drew us to yet another medium: comics, and specifically the work of Scott McCloud. When the artist and author Scott McCloud analyzed the origin of comics, their nature today, and the future of this popular media form, in *Understanding Comics* (1993) and *Reinventing Comics* (2000), he did not produce his analysis as a traditional monograph. He came up instead with the ingenious idea of writing his history and theory of comics in the form of a comic book. He drew himself as a talking character, who delivers—balloon by balloon—a lecture on the subject to which he has devoted his life.

In a conventional book, text and illustrations are separate modes of expression, but in *Understanding Comics* they are intimately connected. Each panel is both text and illustration. When McCloud wants to define how comics function as sequential art, he actually shows the reader how the panels form sequences to represent the passage of time or cause and effect. McCloud as a character is literally drawn into the argument he is making (figure P.1).

McCloud's playful approach inspired us to try to explain a medium in and through the same medium. What McCloud did for comics, we want to try to do for AR and VR. We seek to explain AR and VR in and through AR and VR. But we admit at the outset that we cannot entirely follow McCloud's example here. McCloud was able to write an essay on comics in the form of a comic because both prose essays and comics exist in the same medium of ink on paper. In our case, AR and VR cannot function on the printed page; they are born-digital media forms. It is difficult to illustrate AR and VR in the pages of a printed book, and equally difficult to incorporate a lengthy textual discussion in an AR or VR application.

While McCloud could express himself in a single medium, we need two: print and digital, each favoring a different format, a different ratio of text and image, and different ways of engaging the reader or user. The lack of a single unified medium is a limitation, but it allows us to present the media forms from two perspectives—from the inside and from the outside. The digital version allows the reader to get inside AR and VR—to interact with or even inhabit the media. The printed version is the opposite of an immersive AR and VR experience, contextualizing the technologies that the user experiences in the digital version. It describes the places of AR and VR as media in the history of other media, and it speculates on the future of AR and VR. The printed version is independent of the digital, although the two are in dialogue. A reader will be able to understand the printed book without experiencing the digital version, and vice versa, but your understanding will be enhanced (we hope) if you experience both forms.

There is another major difference between McCloud's project and ours. In *Understanding Comics*, McCloud was presenting a medium with a long and richly varied tradition that he could draw on for evidence. In our case, although they have long been laboratory technologies, AR and VR are just beginning to constitute media with thousands or millions of users. AR and VR are in the process of developing in an ever-changing environment. The

Figure P.1

In Scott McCloud's *Understanding Comics*, text and illustrations are unified into one form of expression. Reprinted with the permission of HarperPerennial, *Understanding Comics: The Invisible Art* by Scott McCloud. © 1993 by Scott McCloud.

American research and advisory firm Gartner issues yearly reports of their so-called hype cycle of emerging technologies. In 2018, they indicated that VR had just moved off the "slope of enlightenment," while AR was still in the "trough of disillusion" and had years to go before reaching maturity (Panetta 2018). As of 2019, Gartner decided that the AR cloud was on the rise, whereas AR and VR were part of technologies that were no longer emerging and would not be included in the hype cycle (Panetta 2019). Gartner's hype cycle is only one indicator of the repeated waxing and waning of these technologies during recent decades. But as many application genres and commercially available devices that support AR and VR have emerged, the impact of AR and VR as media forms is becoming increasingly clear.

Both the printed book and its digital companion are addressed to those who want to understand augmented and virtual reality as media. We consider where AR and VR come from, how they function, and how they have come to take a place among the many media channels and platforms that exist today. This requires some consideration of their technical, as well as historical and cultural, contexts, but neither the printed book nor the digital companion delve deeply into the technologies per se. Those who want to learn how to design and implement AR and VR applications or to seek a thorough treatment of computer graphics (CG), tracking technologies, and computer vision will need to look elsewhere—for instance, to the excellent work of Aukstakalnis (2016); Billinghurst, Clark, and Lee (2015); Hughes et al. (2014); Jerald (2016); Pangilinan, Lukas, and Mohan (2019); Parisi (2015); and Schmalstieg and Höllerer (2016). We limit our discussion of technical details to the ones that are the most important for situating AR and VR in the particular media tradition that we are identifying, the tradition of what we are calling *reality media*. And we strongly believe that both design and technical development can benefit from understanding AR and VR as reality media.

Acknowledgments

Our own thinking about reality media is informed by numerous scholars and computer scientists and researchers, as well as designers, filmmakers, artists, and programmers. Throughout the writing of this book, which culminated in the spring and summer of the COVID-19 pandemic, we met with and discussed our work with many people, and we cannot acknowledge them all individually. Indeed, all our colleagues at the Georgia Institute of Technology and Malmö University have contributed to the environment of discussion and inquiry that nourished our work. Among colleagues in computer science, design, digital studies, and other disciplines at our own institutions and elsewhere, we owe particular thanks to Sean White, Jaron Lanier, Steve Feiner, Mark Billinghurst, Diane Hosfelt, Maribeth Gandy, Joshua Fisher, Rebecca Rouse, Nassim Parvin, Carl DiSalvo, Bo Reimer, Susan Kozel, Temi Odumosu, Per Linde, Johannes Karlsson, Sebastian Bengtegård, Bo Peterson, Brooke Belisle, Paul Roquet, Pepita Hesselberth, Maria Poulaki, Lissa Holloway-Attaway, Lars Bergström, and Anders Sundnes Løvlie. Gunnar Liestøl's pioneering work has inspired us and informed our thinking on augmented reality experience design, especially in the realms of education and cultural heritage. We thank our longtime friend and collaborator Michael Joyce, whose artistic exploration of augmented reality and thoughtful critique of digital media have had a formative influence on the thesis of our book. Scott McCloud inspired us in our attempt to create an uncanny digital double for this book. Others who have played key roles in helping to realize the digital version include Colin Freeman, Joshua Crisp, Gheric Speiginer, Alex LaBarre, and Alex Hill. Our students in the Georgia Tech VIP program (under the direction of Ed Coyle) have helped us test prototypes and refine ideas over the past three years. We thank Carina Ström Hylén and Alex LaBarre for their contributions to the book's graphics.

At the MIT Press, we would like to thank Amy Brand and Terry Ehling for their willingness to undertake a project that is both a traditional book and an experiment in digital publication. Thanks to Noah Springer and Kathleen A. Caruso for their stewardship through the publication process, and to Melinda Rankin for her meticulous and thoughtful editing of the manuscript. We are grateful to the anonymous reviewers whose comments helped us make this a better book. Finally, and especially, we thank Doug Sery, who patiently supported this book through its long gestation period.

Above all, we wish to acknowledge family and friends who have provided much support and inspiration: David Bolter, Philippe Rouchy, Fredrik Engberg, Elizabeth Mynatt, and Markus Klintborg.

Introduction

In January 1896, an audience in a basement room of the Grand Café in Paris attended one of the first public demonstrations of the Lumière brothers' all-in-one camera and projector: the cinématographe. One of the films shown was The Arrival of a Train at La Ciotat Station. *All that happens in the approximately fifty-second film is that a steam engine and several passenger cars pull into the station and glide to a stop. As the train comes closer, some in the audience panic. Imagining that the train is going to burst out of the screen into the room, they leave their seats and rush for the door.*

It is a great story but almost certainly apocryphal. The film historian Tom Gunning doubted that an audience of sophisticated Parisians at the turn of the twentieth century would be fooled into thinking that a train could suddenly materialize and crush them (figure I.1). They were astonished, not fooled, and what astonished them was that moving images could seem so real (Gunning 1986; 2009). They were astonished not by reality, but by a reality medium. Film became one of the most important reality media of the twentieth century, and in some ways, it is a forerunner of virtual reality. Let's reimagine that scene in VR.

Seated in your living room at home, you put on your VR headset and are transported back to that hall in the Grand Café. You can look all around you at the other spectators. You can turn all the way around and see the back of the hall. All of this is rendered in realistic 3-D graphics. The Arrival of a Train at La Ciotat Station *begins to play on the screen. When the train approaches, the shiny engine bursts through the screen into the hall*

Figure I.1
The Arrival of a Train at La Ciotat Station (1896).

*and comes to a stop among scattered chairs. The 3-D spectators jump up
and have managed to get out of the way.*

The scene described above announces the arrival not of a train, but
of a new reality medium (figure I.2). When the engine breaks out of the
screen, it crosses the barrier from one reality medium to another: from two-
dimensional, live-action film to the 3-D graphics of virtual reality.

This book is about virtual reality and its counterpart, augmented reality.
In some ways, virtual and augmented reality are not new. We could argue
that we already live in an augmented reality. If we walk down a city avenue
in the evening, neon signs and electronic billboards abound. Video screens
are everywhere—in offices, restaurants, airports. The cityscape is full of
media that we are invited to read and appreciate as a visual experience. And
media scholar Oliver Grau (2003) has claimed that virtual reality is as old
as the Villa dei Misteri in Pompeii, the Sacred Mount of Varallo in Tuscany,
the tromp l'oeil churches of Baroque Europe, or Robert Barker's panoramic
exhibition hall built in London's Leicester Square in 1793.

Figure I.2
The arrival of a train into virtual reality

Like all previous "new" media, what make augmented reality and virtual reality new are the ways in which they satisfy expectations that our media culture has already had, in some cases for hundreds of years. AR and VR are digital media that depend on the recent developments in smartphones (iPhone, Samsung Galaxy, Google Pixel), headsets (HTC Vive and Oculus Quest), and graphics and other software (Unity and WebGL). They are also products of the desires and intentions of individual makers and their communities (e.g., the open-source community on GitHub), as well as broader cultural preferences and economic motivations and constraints.

When media like these first emerge, they not only define a future path for our media culture but also redefine the past. Once there is VR, the Sacred Mount of Varallo (built beginning in the fifteenth century) and Barker's panoramas (at the end of the eighteenth century) do seem to be precursors. Once there are AR applications for smartphones and tablets, we can look back and see AR in the lights of Broadway in the 1930s. One of the ways to tell that a new digital medium is becoming important is when Hollywood starts to take notice. Throughout the twentieth century, film and television were among the dominant media for our culture as a whole. But by the 1990s various forms of digital media had begun to challenge their status in terms of economics as well as the size and enthusiasm of their audiences. Young people began spending more of their time and money on console and desktop video games. Hollywood responded with films that adopted visual techniques from video game graphics (e.g., *The Matrix*) or were repurposed film versions of video games (e.g., *Mortal Kombat, Tomb Raider, Resident Evil*).

Hollywood also began to worry about another digital technology in the 1990s: virtual reality. Unlike video games, VR was not then a direct economic or cultural threat to film and television. Video games had already become an industry with billions of dollars in revenue, but VR was still a laboratory technology, with research supported by the military and a few start-ups. Although consumer products began to show up in the second half of the 1990s and some game emporiums were offering customers the chance to put on a crude headset for a few minutes of play, VR remained a novelty. The entertainment industry, however, picked up on the hype of evangelists that VR would be the medium to end all media—that VR would take over the role of film and television for entertainment and information. Hollywood responded to the hype with a series of science fiction or fantasy films, such as *The Lawnmower Man* (1992), *Virtuosity* (1995), and *Strange Days* (1995). These films suggested that VR was a danger not to Hollywood itself but rather to the whole fabric of our society precisely because, as the evangelists were claiming, it erased the difference between reality and medium.

But the real VR technologies of the 1990s, and even those of the 2000s, were not remotely close to replacing our visual and physical reality. When a user put on a headset, she would never mistake the elementary graphics she saw for the everyday world, and the so-called latency of the headset's tracking capabilities often made her nauseated after a few minutes of viewing. But in cult classics such as *Strange Days* and *eXistenZ* (1999), and, above all, the blockbuster *The Matrix* (1999) and its two sequels, *The Matrix Reloaded* (2003) and *The Matrix Revolutions* (2003), the VR was flawless—not only visually, but for the other senses as well. In purely perceptual terms, the characters in these films (and we as the audience) could not distinguish being inside the VR from being outside in the world. The point of the Matrix franchise was that the whole human race was being lulled by malevolent machines into believing that the virtual world was, in fact, the real one. Read as allegories, the films were warning us not to abandon the familiar and trusted medium of film, which had brought us great narrative pleasures for decades, for this new and dangerous technology.

AR allows us to extend our physical reality; VR creates for us a different reality.

Although these sci-fi films were depicting a technology that was more fiction than fact, filmmakers did understand something fundamentally important about VR (and, as we shall see, AR). These new digital technologies can indeed do the same cultural work that film and television have done for decades. Like film and television, VR and AR are *reality media*. They place themselves figuratively or physically between us and our perception of the everyday world, and, in this sense, they redefine or construct reality itself.

VR redefines reality as a computer graphic world. When we put on headsets, our everyday worlds are replaced with digital visuals and sounds. We experience alternate visual worlds drawn by the computer. Computer-generated or recorded sound is delivered to each ear. When we turn our heads, sensors inform the computer, which readjusts the images and the direction of the sound accordingly. With an AR device, which can be a headset, glasses, or simply our smartphone or tablet, reality becomes a hybrid. We still see and hear the everyday world, but digital text, imagery, and sometimes sound are overlaid on or blended into that world. The physical and the digital are merged. In short, AR allows us to extend our physical reality. VR creates for us different realities.

What Are Reality Media?

All media (e.g., painting, text, sculpture, photography, film, VR) represent aspects of reality, but some media are of a *second order*, representing reality symbolically. A novel is a second-order media form. It may call up a world for us, but it does so through our engagement with the text on the page and by way of our imagination. We focus in this book on media that fashion a reality for us by appealing to our senses, by inserting a layer of media between us and our perception of the world. Some reality media appeal exclusively to one of our senses or concentrate on a small segment of our perceived reality. A painting, for example, addresses our sense of sight, and it compels us to center our attention on the canvas and to ignore, at least momentarily, what we see beyond the painting's frame. It may draw us into a depicted world, but it does not surround us. Other reality media are expansive, appealing to more than one of our senses and replacing or refashioning the world as a whole. Of these, the two most important in the twentieth century were film and television.

For decades, people viewed films only in a theater, and in this large, darkened space, the audience's attention was naturally focused on the screen in front of them. The dialogue, the background music, and other sounds helped to focus the audience's attention. Although the film was not typically 3-D or truly immersive, the audience could become effectively immersed in the world presented on the screen.

Throughout the second half of the twentieth century, the other dominant reality medium, television, was viewed under very different circumstances: at home in a room that might not be darkened at all. Television viewing was and remains in general more casual. But the way in which television connected viewers to networks of entertainment and information beyond their living rooms, and especially the live broadcasts of news and sports, gave television unique prominence in our media culture. Television created imagined realities, as film did, but it also offered glimpses into the social and physical manifestations of wars, crimes, political events, and natural disasters. The appeal of liveness in broadcast television has carried over into streaming media today, such as Twitch.

As a reality medium, VR more closely resembles cinema, especially as it was originally experienced in a theater. In both cases, you make the decision to step outside your daily life for a period of time and into a fashioned reality. In this sense, VR, like cinema, is a special event. AR functions like screen media that are consumed more distractedly, like a television set that is on in the living room whether someone is focused on the program at the moment or not. Just as video screens are integrated into the lived environments of our homes (and into public spaces such as airports, bars and restaurants, and even elevators and gas pumps), AR integrates digital information into the world we see around us. Like television, AR is, or will soon be, an everyday experience. VR and film are unified reality media, whereas AR and television are hybrids.

VR and AR are among the latest additions to our complex media culture, which still includes everything from books and magazines to paintings, radio, television, and film—all competing for attention and status. In that competition, the producers and promoters of each form promise an experience that is unique and, in some way, better than other media—more compelling, more authentic, truer to life. Although VR and AR promise a new way to construct reality itself, these new media are not going to replace film and television in the foreseeable future. For in addition to competing with each other, media today converge in new patterns of creation and consumption. Films are now viewed not only in theaters, but also on television screens at home, on phones and tablets, and even in VR headsets. We watch television shows at home on a television set, but also on our mobile devices when we are elsewhere—in a café, on a bus or subway train, during our lunch breaks at work, or at home, even though we may have larger television sets or screens.

All four of these reality media (film, television, AR and VR) participate in the process of *remediation* (Bolter and Grusin 1999)—that is, the process of the mutual competition and cooperation among all media at any current cultural moment. Books, film, television, video games, AR, VR, and other media are all part of our current media economy. Our everyday experience of the world is filled with all sorts of media, and it is almost unavoidable for us to think of each medium, new and old, in relation to all the others that surround us. New media borrow from existing media in many ways. For example, 3-D video games borrow camera work from film and television while adding a new kind of interactivity by giving the player control of the camera. VR in turn remediates 3-D console video games with a key new feature: complete visual immersion. The borrowing always involves a sense of competition. Producers working in a new medium are claiming that their medium brings us closer to the real than other, older media. Photography, they claim, gets closer to what things really look like than painting. Film is more realistic than photography because it is dynamic. Video games in VR are better than video games on a flat screen because they are more immersive. As audience and users, we are constantly being asked to compare these media (to vote with our attention and our dollars) in terms of the offer of reality that they make.

We noted that Hollywood has been making movies about VR since the 1990s and that popular conceptions of VR have been formed by such movies more than by the experience of VR itself. Far more people have seen Steven Spielberg's blockbuster *Ready Player One* (2018)—itself a remediation of Ernest Cline's 2011 novel—and experienced VR in a film format than have used an Oculus or Vive headset. The film depicts America in 2045, when virtual reality technologies have become ubiquitous and when many people, such as the teenage protagonist, Wade Watts, want to escape from a physical environment marked by poverty and the effects of climate change into a virtual game world fittingly called OASIS. When Wade puts on a headset and a glove, he plunges into a world of perfect (cinematic) computer graphics and interacts seamlessly in that digital environment. The glove and later a bodysuit even give him tactile feedback (figure I.3).

No current VR game offers anything like this degree of engagement. Richard Rushton (2011) called the experience that film, in general, offers its audience *filmic reality*. Filmic reality, Rushton wrote, is the process by which films become part of how we understand our lives and our world. It does not matter whether these objects could exist in today's physical world; they are real within the context of the film in which they appear. In this spirit, we could say that *Ready Player One* is the filmic reality of virtual reality. This film, along with others from *The Lawnmower Man* to *The Matrix* and beyond, has played an outsized role in shaping our culture's perception of the potentials and the dangers of the actual technology of VR in our world. Film has helped to create unrealistic expectations people have of VR; at the same time, the makers of VR systems have used the representational power of film itself as a measure of the promise of VR.

Since at least the 1980s, the creators of VR technology have been telling us (and themselves) that the goal of VR is to duplicate what it is like to see, to hear, and even to feel the world around us. When reality media claim to bring us reality, to bring us life, that claim is always made as an implied comparison with other, usually older media. When we put on a VR headset, we implicitly compare our experiences to film, television, and screen-based video games. VR is felt (by some at least) to be more real than watching a movie because it can give us a 360-degree immersive experience. We are not limited to the particular view that the director wants to show us at any given moment.

Another way to express the difference between film and VR is that film records a world and VR simulates a world with which the player can interact. A VR experience is a computer program that defines a world with visual

(a)

(b)

Figure I.3
Ready Player One as the ultimate VR world. (a) The top image shows the technology in the real world; (b) the bottom image shows the VR world. Licensed by Warner Bros. Entertainment Inc. All rights reserved.

and auditory qualities, its own set of "physical" laws, and modes of inter-action. That virtual world may be a 3-D replica of some part of our physical world. We might take part in a virtual meeting with avatars that look like our colleagues in a realistic conference room. But this is still a different world, an "other" world, when we experience it wearing a VR headset. VR remediates 3-D video games, which have been played on a computer or video monitor since the 1990s and are obvious candidates for VR, precisely because they too simulate a world rather than recording it.

Immersive 3-D graphics in VR and interactivity in both VR and AR are qualities that make them putatively better reality media than film and television. Similarly, when film and television were new media in the first half of the twentieth century, they were tacitly, if not explicitly, offered as improvements to earlier media, such as photography, painting, and stage drama. Film was like photography, only better, because it gave the viewer not only a faithful reproduction of what things looked like but recorded motion as well. A photograph recorded a moment in time, but a film recorded time itself, capturing a further dimension of reality. Television not only captured the dimension of time, like film, but it could be *live*, a broadcast that occurred *in real time* (as we say now).

Reality media do not get us closer to some transparent presentation of the real; instead, they work by comparison. When the screening of *La Ciotat* announced film as a new reality medium in 1896, the audience was astonished for two related reasons. First, they were comparing what they saw on the screen to their experience of everyday life, for which film offered them a new, uncanny double. The train that threatened to crash through the screen was ultimately "a train of shadows," as Russian novelist Maxim Gorky put it (Gunning 2009, 822). At the same time, the audience was comparing what they saw on the screen to other media that they knew, especially photography. The invitation to such a comparison is clear from Gorky's (1896) description of a showing of the film in Nizhny Novgorod, Russia, in the summer of 1896. When the lights went out, a still image of a Paris street was first projected on the screen. Then, "suddenly a strange flicker passes across the screen and the picture comes to life." As Gunning pointed out, starting with a still (photographic) image, which the audience had seen many times before, and then turning it literally into a moving image would not fool the audience into imagining what they were seeing was really there in front of them. Instead, it emphasized the remediation, saying in effect, "you see, this new medium is like photography, only better." From that moment on, reality was redefined in the sense that a new reality medium now existed that gave twentieth-century viewers a new perspective on the real. We will call this the *La Ciotat effect*, and it is the effect that each reality medium has. Each reality medium mediates and remediates. It offers a new representation of the world that we implicitly compare to our experience of the world in itself, but also through other media.

Each reality medium mediates and remediates. It offers a new representation of the world that we implicitly compare to our experience
of the world in itself *and* through other media.

Two Versions: *Reality Media* and *RealityMedia*

This book comes in two versions: one in print and one digital. The printed
version is the one you are reading now. In addition to this introduction,
there are ten chapters. Chapters 1–5 lay technical, cultural, and historical
foundations for understanding the roles of AR and VR in our changing
media culture, and chapter 6 surveys their current applications. Chapters
7–10 outline their potential roles in the coming decades. Chapter 1 describes
some of the key differences between "classic" VR and AR. It introduces
extended reality (XR) and the immersive web and goes on to show where
AR and VR fit on Milgram and Kishino's virtuality continuum. In chapter 2, we sketch a history of earlier reality media, from Renaissance painting to more immediate forerunners of AR and VR, including 360-degree
video. Chapter 3 shows how 3-D computer graphics help to construct the
visual realities of AR and VR. It focuses in particular on photorealism and
the uncanny. Chapter 4 explains how AR and VR's tracking and sensing
technologies contribute to the spatial aesthetics of these media. Chapter 5
further explores their aesthetics by considering how they generate a sense
of presence and aura in their users. In chapter 6, we survey the genres of
VR and AR experiences, including games, art, cultural heritage, training,
navigation, and virtual conferencing. Chapter 7 shows how AR may eventually fashion uncanny mirror worlds of our lived environment. Chapter
8 discusses the vision and the practicality of VR metaverses, virtual worlds
shared by thousands or millions of users. Chapter 9 examines the ways
in which VR and AR join earlier digital media in redefining the relationship between public and private space. We explore the implications for the
future of privacy. Finally, in chapter 10, we speculate on the future of AR
and VR as media: what is possible, plausible, and probable as they become
increasingly integrated into our media environment. We revisit the genres
of chapter 6 and the mirror worlds and metaverses of chapters 7 and 8.

Chapters 3 and 4 are the most technical chapters, but they are not text-
book discussions of computer graphics, tracking, and sensing, nor do they

discuss the latest research in any systematic or scientific way. We focus instead on a few key techniques and technologies that make AR and VR distinctive as media. Chapters 2 and 5 include the most relevant earlier media, such as photography and film, and theoretical concepts, such as presence and aura. The long chapter 6 on the genres of AR and VR can be sampled. As we acknowledge at the beginning of that chapter, many of the examples will soon be out of date, although we believe that the genres themselves will prove much more long-lasting.

The digital version of this book, *RealityMedia*, serves as a complement to, not a substitute for, the one in print. Its goal is to allow you to experience the reality media that we describe in print—to explain AR and VR in and through AR and VR. *RealityMedia* consists of web pages and a set of virtual rooms that correspond to the chapters of the printed book. Within each room, you will find 3-D models, videos, images, and texts related to the themes and issues raised in the corresponding chapter. In some of the rooms, you will be invited to interact with the models; other 3-D models or videos simply run and invite you to watch.

You access the digital *RealityMedia* at https://realitymedia.digital via the browser on your computer, iOS or Android device, or VR or AR headset. The experience will be somewhat different depending on the capabilities of the device you use. For example, you will not have a fully immersive experience on your computer console or the iPhone. You will find a list of devices and browsers with which the book can currently be used at https://realitymedia.digital/devices. No matter what device you use, you will begin on an introductory web page with text and embedded images. Clicking on an image will open up a corresponding room to explore.

RealityMedia is a remediation of *Reality Media*, but it is not a reproduction. It contains materials that could not be included in the print version, often because it is clearer and more compelling to demonstrate or enact features of AR and VR than to describe them in words or illustrate them in static images. It also provides (we hope) a different and valuable perspective on our argument.

1 What Are Augmented Reality and Virtual Reality?

Even a few years ago, relatively few people had experienced augmented reality or virtual reality because the required hardware was expensive and the software difficult to use. Now several million VR headsets are shipped each year from various manufacturers—Sony, Oculus, HTC, and others (Statista, n.d.). Many gamers have experienced VR versions of role-playing games (RPGs), construction games, driving games, sports games, or first-person shooters (FPSs) such as Valve's *Half-Life: Alyx* (figure 1.1).

Valve's *Half-Life: Alyx*, a first-person shooter game in VR, was the long-awaited continuation of one of the most successful desktop FPS series. The first *Half-Life* game was released in 1998 and set a standard that other FPS

Figure 1.1
Valve's *Half Life Alyx*. A first-person shooter game in VR. ©2019 Valve Corporation. Reprinted with permission.

games followed. Like earlier titles in the series, *Alyx* is a sci-fi horror fantasy that takes place in a wasted city under attack by multidimensional aliens. Released in 2020, *Alyx* received enthusiastic reviews, which is unusual for the often-critical hardcore gaming community, and set a standard for VR gaming (Fuscaldo 2019; McKeand 2020; Metacritic 2020). You play Alyx Vance, a young woman who is part of the resistance against the alien invaders. You wander through the ruins of City 17, solving puzzles and killing a variety of animate and robotic enemies. Shooters like *Alyx* have long used 3-D graphics to create an immersive experience for millions of players. And for decades, players on computers and game consoles have yearned for true VR so that they could fall through the screen into the worlds on the other side. The shooter game genre is easy to remediate in VR—easy conceptually, that is, although the design and programming present special challenges (Reeves 2020). *Alyx* was designed to take full advantage of what the available consumer-level VR headsets (such as HTC Vive and Microsoft Mixed Reality devices) offer as a platform; there is no non-VR desktop version.

As for AR, headsets or glasses, such as HoloLens 2, Magic Leap, and the Vuzix line, are currently more expensive than their VR counterparts. However, billions of users throughout the world own smartphones that can deliver screen-based AR experiences. The AR game *Minecraft Earth*, for example, an addition to the extremely popular Minecraft sandbox games, was designed for smartphones. Other versions are played on a variety of computers or game consoles, and there is a *Minecraft VR* as well, available for some headsets. In all these earlier versions, the Minecraft world is entirely virtual, rendered in computer graphics, but as the name suggests, *Minecraft Earth* takes place in the physical world all around the player, either indoors or outdoors. The player first collects playful animals and architectural elements on her screen, all with the same blocky cartoon aesthetic as in the other versions of the game. She assembles these objects on "build platforms," which she can then set down in her living room or backyard or in the local park. She can scale these platforms to fit on a tabletop or to be as large as a small house outside. With other players, she can manage these structures, adding and interacting with animals or changing the architecture. And each structure can persist; she can come back the next day and engage with it again. In the last decade, *Minecraft* has gathered a community of millions of players and produced several spin-off games and elaborations to the basic work of construction, involving adventures and combats, but always maintaining a benign, child-friendly aesthetic, unlike FPS games such as *Half-Life: Alyx*. In January 2021, Minecraft announced that it would shut down *Minecraft Earth* in June. Nevertheless, the game exemplifies the hybrid character of AR, integrating the digital and physical worlds, and will likely be a model for creative AR in the future.

Games such as *Half-Life: Alyx* and *Minecraft Earth* show what VR and AR offer today for consumers. Let's consider briefly how the two technologies got to this point in their development.

The Origins of AR and VR

The development of AR and VR as computer-based technologies is generally said to begin with the work of the graphics pioneer Ivan Sutherland, who constructed the first responsive, head-mounted system in the late 1960s (Sutherland 1968). In one of its modes, Sutherland's headset was suspended from the ceiling by a mechanical boom in order to track the user's head movements, leading to its nickname: the Sword of Damocles. The system displayed a wireframe object floating in the room in front of the user. It relied on *see-through* optics, which employed half-silvered mirrors to enable the user to see the physical room itself, as well as the virtual object that the computer was drawing. When the user turned her head, the object remained in place. The Sword of Damocles is often considered the first VR headset, but today we would more accurately call it AR. At that time, the technology (both the hardware and the software) was not capable of constructing a complete graphic world. VR was out of the question, although it was certainly one of the goals.

> **AR and VR were twins at birth, beginning as variations of the same technological idea.**

The years that followed were the heroic age of computer graphics, when many of the fundamental techniques were invented—in some cases, by Sutherland's own students. For example, Ed Catmull, who received his PhD under Sutherland at the University of Utah, invented methods for digitally representing curved surfaces and the idea of *texture-mapping*, a technique for covering 3-D objects with detailed 2-D patterns to make their appearance more realistic. Nevertheless, computer graphics was a minor part of the burgeoning field of computer science in the 1970s. Most computer scientists thought there were far more important areas to study (such as programming languages, operating systems, and database design) and more important problems to solve than how to draw pictures effectively on cathode-ray tubes. When the premier academic conference for graphics, SIGGRAPH, first convened in 1974, there were only six hundred attendees (ACM SIGGRAPH, n.d.).

Two developments in the late 1970s and 1980s brought computer graphics to the attention of both computer scientists and the general public. One was the graphical user interface (GUI), invented at the Xerox PARC lab and featured on the Apple Macintosh in 1984. The other was computerized special effects in film. In 1977, the space epic *Star Wars* was an unexpected success, and George Lucas used some of the profits to set up a computer division of Lucasfilm, which employed Catmull and Alvy Ray Smith, a young PhD from Stanford. Catmull and Smith went on to found the animation house Pixar in 1986, funded by Steve Jobs (Pixar, n.d.). Pixar combined 3-D graphics and clever storytelling to produce a remarkable line of films, from *Toy Story* (1995) to *Incredibles 2* (2018). So while Apple computers were putting graphics on the small screens in front of a growing number of personal computer users, Lucas and other Hollywood producers were impressing millions with computer-generated imagery (CGI) effects on theater screens. The growing commercial importance of graphics led more computer scientists to work in the field, leading in turn to better software and hardware. In 1987, there were over thirty thousand attendees at SIGGRAPH, a figure that included participants at a trade show, with representatives from industry outnumbering scientists and engineers.

Improvements in computing power, display technology, and software meant that true VR and AR were finally becoming possible. VR was pursued sooner than AR and attracted far more publicity. One reason was probably that VR appealed more powerfully to the imagination of technologists, enthusiasts, and the general public. VR has been called the ultimate medium, one that replaces the mundane physical world and transports viewers to any other world of their choosing. In that ultimate form, VR seemed to many to challenge reality itself—a technology so perfect that it becomes invisible, a reality medium that is no longer a medium.

In the 1960s, Ivan Sutherland himself had fantasized about the ultimate computer display, which would allow the user to fall through the screen into a graphic and interactive digital world (Sutherland [1965] 2009). And in the 1980s, Alvy Ray Smith is reputed to have said that "reality is 80 million polygons per second" (Rheingold 1991, 168). Research accelerated in the 1980s—for example, at NASA's Ames Research center, under the direction of Scott Fisher. In this period too, Jaron Lanier became one of the chief mythmakers for VR, and he continues to be an articulate spokesperson for the vision of VR as fashioning alternative digital worlds (2017). Lanier popularized, though he did not coin, the term *virtual reality*. And in 1984, he founded VPL Research, one of the first companies to build and market VR systems. VPL Research pioneered systems and techniques, while Lanier was trying to sell VR to the military, venture capitalists, and Hollywood (Sorene 2014). VPL's Eyephone headset presented a polished version of the technology that was well-suited to Hollywood's notion of VR as a sci-fi technology in the *Star Wars* decade. If one were to compare the Eyephone to Sutherland's Sword of Damocles, it would show not only how far the technology had progressed but also how it was becoming oriented to a world of entertainment and popular consumption.

In 1990, VPL's equipment helped to promote Hollywood's representation of this new reality medium by serving as props in the early dystopian film *The Lawnmower Man*. Lanier's own view was and remains anything but dystopian: that VR could usher in an era of creativity and "postsymbolic communication" among humans in computer graphic worlds. Who needs language and symbols, he seems to suggest, in a virtual world where you can make 3-D models of anything and everything? Lanier does not, however, subscribe to the hacker fantasy that we can or should leave the physical world behind and live exclusively in VR. He has always focused instead on the aesthetics of VR, regarding VR as a medium that can sharpen our appreciation of the physical world. He writes: "VR still teaches me. I love noticing my own nervous system in operation, and that's more possible in VR than in any other circumstance. I love seeing nuances I used to be blind to in the light and motion of the natural world, in forest leaves and in children's skin. This happens most intensely when you compare reality and VR" (2017, 285). This is exactly what we mean by a reality medium: comparing our experience of our everyday world with VR inevitably redefines the reality of that experience.

Throughout the 1990s and the 2000s, Hollywood continued to make films about VR, produced with computer graphic special effects but displayed to audiences on flat screens. In the same era, the burgeoning video game industry tried to find ways to market a true VR game. Three-dimensional graphics were tremendously successful in first-person shooters and role-playing games on TV screens and computer monitors. It seemed obvious that a VR headset would offer the players (at this time, largely teenage boys and young men) the ultimate immersive experience they craved. But products such as the Sega VR-1 (DigitalNeohuman 2010) and Nintendo's Virtual Boy failed commercially because 1990s technology could not provide attractive graphics and responsive gameplay at a reasonable price. Serious research on VR continued in labs, often funded by the military, but high-quality VR equipment remained too costly, too bulky, and too complicated outside of the lab, even for most high-end commercial uses.

In 1992, a very different VR platform, called the CAVE, was developed by Carolina Cruz-Neira and colleagues at the University of Illinois (Cruz-Neira, Sandin, and DeFanti 1993). A CAVE dispensed with bulky headsets and the need for the user to be tethered to a computer. Instead of drawing images on two small displays that the user wears, the CAVE projected large images on four walls, and sometimes on the ceiling and floor, so that the user was literally surrounded by them. But a CAVE could never be a general interface for millions of users because it was quite expensive and had to be set up in a dedicated space. So ironically, while experts were laboring to bring VR beyond the stage of blocky graphics and equipment that cost tens of thousands of dollars, Hollywood films were warning audiences about the danger of falling into a perfect VR world that the current technology could not possibly create.

By the end of the first decade of the 2000s, however, every aspect of VR (graphic displays, tracking, and software) had improved, making possible some high-end commercial and military applications—for example, simulating the cockpit of a fighter jet or a tank for training. Then, the development of the Oculus Rift starting around 2011 dramatically reduced the cost of good-quality headsets and enabled VR to become a platform for games and other applications (Wikipedia contributors 2020g). Other VR headsets (HTC Vive and Sony PlayStation VR) costing hundreds rather than thousands of dollars followed, and VR became a viable reality medium. Today's headsets are sleeker, but their form factor is not that different from VPL's gear of the late 1980s.

In contrast to VR, AR after the Sword of Damocles did not develop as an independent research area until the 1990s. The foundational work in computer graphics would eventually benefit AR as much as VR, but there were technical reasons that AR lagged. AR requires precise tracking, even more so than VR does. For many AR applications (especially indoor apps like placing virtual furniture in a room or playing a desktop game), the system needs to know exactly where the user is looking in order to line up graphical images with her view of the physical world. That degree of precision was hardly possible until the 1990s. Furthermore, many of the applications envisioned for AR require the user to walk or move around freely, and in the 1990s it was not possible to build sufficient computing power into a portable headset. The AR experiments in labs were often conducted with systems tethered to stationary computers and trackers wired into the ceiling.

Beyond the technical obstacles, there was a cultural one: AR does not lend itself as easily as VR to utopian or dystopian mythmaking. Jaron Lanier was suggesting that VR would enable children to play games inside the head of a giant, friendly, amethyst octopus (2017, 89), and Hollywood was imagining that VR would take over (and possibly destroy) our lives. Compared to these visions, an AR application that would allow you to try on a dress before buying it did not seem exciting or threatening. AR researchers proposed practical commercial and military applications, such as repairing machinery or providing a heads-up display for pilots. Tom Furness's HITLab at the University of Washington and Steve Feiner's lab at Columbia University exemplified such AR research in this period, as did experimental work at Boeing, where David Mizell and Thomas Caudell tested the use of AR to help workers assemble the wire bundles for a 747. Caudell is reputed to have coined the term *augmented reality* (Lee 2012).

The AR community continued to work on the hard problems of tracking through the first decade of the 2000s, but without much publicity. Then, about the time that the Oculus Rift vastly expanded the market for VR, the development of powerful smartphones began to make AR widely accessible. Many AR researchers were not enthusiastic about smartphones, believing that true AR required a see-through, head-worn display—something like Microsoft's HoloLens 2, which allows for a continuous integrated visual experience. But accepting the smartphone as a platform for AR meant millions of potential users, who already owned phones for a dozen other purposes. Today there are a range of platforms for AR, from phones to modified glasses to full headsets equipped with cameras, all contributing to the confusion that characterizes this medium not only for the audience but for developers as well.

The Virtuality Continuum

AR and VR have developed for decades along separate paths, and AR and VR headsets remain fundamentally different. Compare the Oculus Quest with the Magic Leap 1 (figure 1.2). The Magic Leap looks like a pair of goggles; the user who puts on the Oculus Quest essentially has a box in front of her eyes.

Although both systems rely on computer graphics to draw virtual objects, the technologies for displaying them differ significantly. VR has

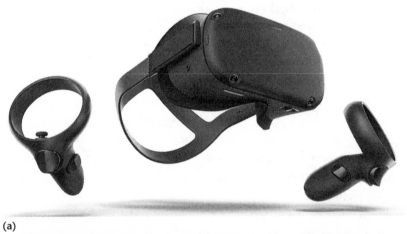

(a)

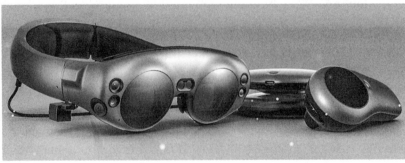

(b)

Figure 1.2
(a) The Oculus Quest is a VR headset © Facebook Technologies LLC.; (b) the Magic Leap is AR © Magic Leap, Inc.

two sets of optics, one for each eye, that together produce a wide continuous field of view (often about 110 degrees at present). An AR device uses one of two techniques for combining the view of the physical world with what the computer draws: video-mix or see-through optics. Video-mix optics takes a video of the world and mixes the computer graphics in. This is how smartphones do it as they have one or more cameras on the back anyway. See-through optics is the more advanced method, used in HoloLens 2 and Magic Leap. The natural light from the world passes through a lens in the goggles, and the lens system itself adds the digital information to the view.

Both forms of AR optics are distinct from the optics of VR, and this technological difference, in turn, reflects the key aesthetic difference between these two reality media. VR surrounds and isolates the user in a computer graphic world. Everything the user can see is drawn by the computer. AR replaces or overlays only a part of the physical world; the rest is still visible, at least as seen through a video camera. Where VR replaces the world with another, AR keeps us situated in the world. With this fundamental distinction, classic VR and AR would seem to be suited to different kinds of applications. VR is ideal for immersive role-playing games, such as *World of Warcraft*, where the player is transported to a fantasyland of pseudomedieval fortresses, towns, and forests. AR is far better suited to a tourist application to help the user find her way through an unfamiliar city and identify restaurants, hotels, and sights. Despite these differences, however, AR and VR share so much hardware and software that it is almost inconceivable that only one of them would have been developed without the other. And today, if an AR system uses a video camera to display the physical world, as smartphones do, then it is at least possible to switch seamlessly between VR, in which the computer draws everything, and AR, in which the camera provides a view of the world presented as background video and everything else in the AR space is drawn in computer graphics.

The relationship between AR and VR can be illustrated with a portal. AR portals became popular in 2017 and 2018 to demonstrate the power of Apple's ARKit and Google's ARCore, which made it much easier for phones to display stable virtual objects in the user's view of the world. If the user is outside in the street, a virtual object—for example, a doorway—can be placed in front of her, and she can walk around it or through it. There can be a different scene on each side of this portal. One side can be AR, in which the portal appears anchored in the street where she is standing. The other can be VR, created entirely in computer graphics. When she steps through the portal from the AR side, she enters a VR world to explore. If she turns around, she can see that this VR world has a portal that leads back into the physical world (figure 1.3). As this portal suggests, AR and VR occupy a continuum.

Figure 1.3
A simulation of an AR portal in the physical world. Stepping through the portal leads into a VR world. Graphics by Alex LaBarre.

In the 1990s, when both VR and AR were still laboratory technologies, two researchers, Paul Milgram and Fumio Kishino, diagrammed what they called the *virtuality continuum,* a spectrum with the *real environment* (the physical world) at one end and the complete *virtual environment* (VR) at the other (Milgram and Kishino 1994; Milgram et al. 1995). They placed augmented reality and augmented virtuality as intermediate points between these endpoints. They used the term *augmented virtuality* to describe a system in which most of the environment is drawn by the computer, but some objects or people from the physical world bleed through into the virtual. Far more common, then as now, were systems for augmented reality, in which most of what the viewer sees is the physical world, overlaid with some computer graphic objects. Their umbrella term, *mixed reality,* referred to any point in the middle, any hybrid of the physical and virtual (figure 1.4).

In recent decades, many of these hybrids have been realized—some as audio-only experiences, some featuring text and images, some playing live video, some displaying 3-D graphics, some requiring precise registration (as with the portal), some not. Not all researchers agree on which of these should be called AR or VR, although users are less concerned with definitions than with the experiences that their computers and smartphones provide. Today, people in the developed economies of North America, Europe, and Asia spend much of their time in a mixed reality toward the left end of the virtuality continuum, between the real environment and "true" AR. They do not wear headsets, but they work in offices surrounded by computer monitors and spend their free time staring at the world around and through the screens of their smartphones. Their reality is an almost constant mixture of the physical and virtual.

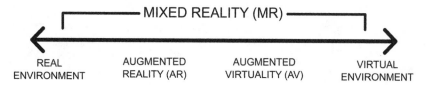

Figure 1.4
Milgram and Kishino's virtual continuum joins AR and VR (Milgram et al. 1995). Reprinted with permission.

The Milgram and Kishino continuum can be illustrated in our example portal. If you are on the street side, you are in AR. You can glance into the computer graphic world beyond the portal, but most of the scene is the physical world. If you closed the portal entirely, then you would be in the physical world, at the left end of the spectrum. If you open the portal, enter the virtual side, and then look back, the portal gives you a view of the physical world as if it were embedded in the virtual world. This would be augmented virtuality. Close the portal down completely while you are on the VR side, and you are in pure virtual reality (figure 1.5).

Researchers still frequently cite Milgram and Kishino's virtuality continuum as a way of understanding the range of reality media technologies. But in 1995, Milgram and his colleagues went on to offer three more scales for assessing mixed and virtual reality systems. These three are seldom mentioned today, although they in fact raise issues that are still important for understanding reality media. The first is *reproduction fidelity*: How well does the system reproduce the virtual and physical objects the user sees? VR and to a large extent AR rely on computer graphics to achieve reproduction fidelity, as we discuss in chapter 3. The second is the *extent of world knowledge*: How much does the system know about the physical world that the user is looking at? What an AR and to some extent a VR system can learn about the user and the world around her is determined through techniques of tracking and sensing, as discussed in chapter 4. Finally, there is the *extent*

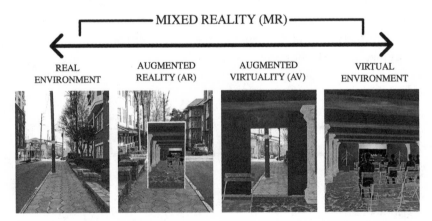

Figure 1.5
The portal illustrates all four points on the Milgram and Kishino continuum.

of presence metaphor: To what degree is the user meant to feel "present" in the scene that she sees around her? This notion of presence remains the holy grail of VR, although we will argue in chapter 5 that it needs to be understood not in the naive terms of the La Ciotat myth, but rather in the more sophisticated way that Gunning attributes to viewers of those first films.

XR and the Immersive Web

The software and the hardware for VR and AR continue to evolve and converge. New phones include multiple cameras or even depth cameras, which together with computer vision software make surface detection and eventually scene understanding possible. There are inexpensive stand-alone VR headsets, as well as enclosures that allow you to convert your phone into a makeshift VR device. Some VR headsets are equipped with forward-looking cameras, which makes them capable of video-mixed AR, although a high-quality multipurpose AR-VR headset still lies far in the future. Even without a headset, the same app on a smartphone may jump between AR and a form of "pan-and-scan" VR. The Milgram and Kishino reality-virtuality continuum accommodates this convergence. The authors listed only four principal points—the real environment and three technologies (AR, AV, and VR)—but they described a host of configurations of visual displays even in the 1990s.

Today the term *extended reality* (XR) is often used to describe the technologies along this continuum including AR, VR, and other forms of mixed reality. Whether running on a headset or on a smartphone, most XR today takes the form of an individual application programmed or adapted especially for one family of devices, such as the iPhone, the Android, the Oculus Quest, the HTC Vive, and so on. But a universal platform could reach many more users, and such a platform exists: the World Wide Web. WebXR is a standard designed to allow web browsers to display AR and VR experiences just as they display web pages now. The web is the conduit for so much information (thousands of petabytes; google *petabyte* and count the zeros) precisely because it is so widely available, and now AR and VR can share in that ubiquity. Every smartphone, tablet, laptop, and desktop computer offers some version of the web browser, usually several versions. Most of the headsets do as well. Many standard browsers are being adapted to handle the WebXR extension, and these can become portals to an immersive web. In the future, an immersive web application will only need to be written using the programming interfaces defined by WebXR, and it should then run on practically every device.

The language of WebXR also allows applications to reconfigure themselves in interesting ways to accommodate different forms and devices. An application can launch as a web page, with text and images that you read like any other page until you come to a link. Clicking on the link could open up a VR world. If you are on a laptop, the VR world would appear on the screen as a 3-D space. If you have a stand-alone headset or one attached to your computer, you might now put it on and begin a fully immersive experience. Or the page might remain a hybrid, with the VR world visible in only a portion of the screen, like a YouTube video embedded in an ordinary web page. WebXR invites a new kind of thinking about information design in which pages can begin in two dimensions, morph into 3-D, and then flatten into two dimensions again.

Applications of the two reality media expand when they become part of the immersive web. The web is a lightweight platform for information. Most of us open our browsers dozens of times each day, so AR and VR applications on the web have the potential to invite us for short, opportunistic visits. This is especially true of AR and VR experiences that do not require headsets, although in the future putting on the headset may not seem any more troublesome than putting on headphones to listen to music (chapter 10).

AR and VR and the Senses

Around 2015, both VR and AR entered a new phase, in which the connection between the physical and virtual worlds is becoming both deeper and broader. The visual connection is becoming deeper because of advances in *computer vision*. The world the computer sees through its camera(s) is no longer always flat. Desktop computers, headsets, and even smartphones running computer vision algorithms are now beginning to see *into* the world: to detect surfaces and identify the shape and volume of objects in space. And detecting surfaces, such as the floor, tables, and walls, is a first step to understanding the scene as a whole. Complete scene understanding means that tables, chairs, animals, and people are no longer just pixels on the surface of the camera's sensor chip. Computer vision algorithms combine that data with data from the smartphone's or headset's motion to analyze the tables, chairs, and people as three-dimensional objects and to organize those objects into a scene.

For AR, surface detection and scene understanding are proving to be nothing short of a revolution. Surface detection promises that AR apps running on devices available to millions of users will finally be able to live up to popular expectations. It is now possible to stabilize virtual objects in the user's field of view, as we saw in the portal that the user can step through (figure 1.3). When the user takes those steps, the portal must behave like an object in the real world and remain in place. Going beyond surface detection, scene understanding promises to reach the point where the system could eventually recognize all the objects in a room and their spatial locations. Imagine that your phone could identify all the furniture, plants, animals, and people that the camera sees around you. The potential applications are vast and will at the same time raise serious issues of privacy.

Prior to the development of these techniques, there have been other methods for registering objects in the world, using the Global Positioning System (GPS), for example. The 2016 version of the first truly successful AR game, *Pokémon Go*, relied completely on latitude and longitude as conveyed by the GPS receiver in the player's phone, and this information could be inaccurate by ten meters or more. This meant that a PokéStop that was supposed to be located in a park could appear in the park to one player and across the street to another. Designers of such outdoor, location-based games, and indeed any practical AR applications, had to take that inaccuracy into account. With GPS alone, smartphones could never achieve the accuracy or consistency that would be needed for a whole range of applications. Now, however, the reality of AR is beginning to catch up with the cultural myth.

In this respect, AR reminds us of several other recent technologies that have taken on the status of a cultural myth soon after (and sometimes before) they were developed. Such myths are often born in science fiction novels and movies, which are free to imagine the technology as a fully functioning part of our material world. This was true of lasers (death rays in decades of science fiction, pulps, novels, serials, and films), artificial intelligence (HAL from *2001: A Space Odyssey* [Kubrick 1968] and a host of similar malevolent computers), and virtual reality (*The Matrix*). AR began to be imagined in movies as early as the first *Terminator* (1984) and later in novels such as William Gibson's *Virtual Light* (1993) and Vernor Vinge's *Rainbow's End* (2006). These were fantasies at the time, but now highly accurate registration through computer vision is permitting AR to live up to at least some aspects of the myth.

Computer vision can also extend and refine VR, although not to the same degree as AR. VR entertainment experiences, especially games, do not benefit in the same way from computer vision techniques because they do not need much, if any, connection to the user's physical world. If the player of *Half-Life: Alyx* is seated in a physical chair, she can walk or ride on a subway car through City 17 without moving from that chair. Because the space of the physical world has effectively disappeared from this construction of reality, no scene understanding is needed. Even if the player chooses to move physically in the game experience, which is possible with *Alyx*, the typical solution is to clear an area in a room, so that there is nothing physical to collide with. Improved scene understanding can lead to VR experiences that exploit the physical space around the player or user for hybrid games. If scene understanding were perfected, you might even be able to wear your VR headset while walking down a city street without bumping into passersby or falling into traffic. This may be part of VR's future in a few decades (chapter 10).

The connection between the virtual and the physical in AR and VR becomes broader when senses beyond vision can be engaged more directly than before. There is growing attention to the way that sound can define a virtual space in VR and enhance the experience in AR. In recent years, audio-only applications have become common. It is now possible to represent spatial sound in greater complexity, to populate an aural landscape with specific sound sources. Distant sources actually sound far away and can become louder and more present as the user gets closer to them. Audio has of course been a key element in the earlier reality medium of film since the late 1920s. For decades, however, most sound in film was flat. Eventually, stereo sound was introduced into theaters, but the spatial effect was limited. At the end of the twentieth century and the beginning of the twenty-first, more elaborate sound systems in theaters (and home theater systems) have made spatializing and depth effects familiar to audiences. But those audiences are still seated, so the sounds must come to them. With

AR and VR, users can move toward sounds or away from them, which lends the space an added sense of depth and texture.

Two other senses are also playing an increasingly important role in AR and VR: touch and *proprioception* (the sense of spatial orientation and being in the world that we gain from the movement of our bodies). Controllers for VR headsets enable users to select objects, type responses, and navigate through virtual spaces. They are integral to VR games such as *Half-Life: Alyx*. Both touch and proprioception are brought into play in Google's *Tilt Brush* (figure 1.6), which illustrates how VR can define space through interaction. Holding a hand controller as a brush, the user can paint with light, and the objects she creates persist in the space around her. In fact, all the VR games that use hand controllers are tactile experiences.

The Aesthetics of VR and AR

Beginning as early as the 1960s, artists played a role in mapping the dimensions of virtual, augmented, and mixed reality, even if they did not

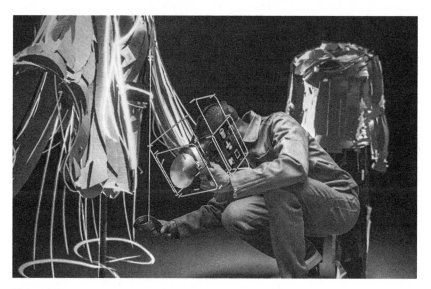

Figure 1.6
Tilt Brush, for painting in virtual 3-D space. This composite image shows the user as if he were in the space that he is painting. Users do not see themselves this way. © *Tilt Brush* by Google 2016. Reprinted with permission.

contribute directly to their technical development. They partnered with technologists to imagine new forms of expressive media and explored the affordances and aesthetics of what we could later recognize as AR or VR. In 1962, the cinematographer Morton Heilig patented his Sensorama. Combining a 3-D projection system with stereo sound, vibration, wind effects, and even scents (Loeffler 2019), the device sought to create an immersive experience—for example, of driving a motorcycle through streets in Brooklyn, feeling the vibration of the machine and the wind on one's face. The Sensorama aspired to provide a totally mediated environment and in that sense anticipated the ambition of VR. But as it simply strapped together earlier technologies rather than unifying around a coherent technology, Heilig's Sensorama did not suggest a viable path forward as a medium, and only a few copies of the device were ever made.

In 1967, Experiments in Art and Technology (E.A.T.) was founded by engineer Billy Klüver, featuring artists as well-known as Robert Rauschenberg and John Cage, and some of these experiments could be seen as mixed or augmented reality. Computer scientist and pioneer digital artist Myron Krueger established Videoplace at the University of Wisconsin in Madison—a lab for what he called *artificial reality*. His interactive installations used projection rather than a headset and could be regarded as forerunners of the CAVE of the 1990s. The Ars Electronica festival also was established in Linz, Austria, in 1979 and since then has exhibited indoor and outdoor installations in mixed, virtual, or augmented reality. In earlier decades, the expense and complexity of AR and VR equipment were limiting factors, but today inexpensive headsets and smartphones and simpler software are also making for a renaissance of digital AR and VR art.

Digital artists have thus been contributing for decades to the aesthetics of AR and VR. But there is a definition of *aesthetics* that encompasses a much broader range of human experience: aesthetics as "relating to perception by the senses; received by the senses" (OED Online 2020). This broad definition was what the media pioneer Marshall McLuhan had in mind when, in *Understanding Media* (1964), he argued that media extend our "sensorium" and that each medium does so in a different way. Admittedly, McLuhan's concept of a medium was one-dimensional. He considered only the technological characteristics themselves (film was photographic images displayed at twenty-four frames per second; TV was an image on a cathode ray tube with 486 visible scan lines). He regarded the social impact of each medium as secondary, the product of the technical characteristics, when in fact the social, cultural, economic, and technological aspects of a medium are all interrelated. Nevertheless, McLuhan's fundamental insight is worth keeping in mind. Media do filter, focus, or amplify our sensorium. Some isolate one of our senses; others address multiple senses. A microscope allows us to see objects that are too small for our unaided eyes but limits our vision to a tiny patch of the world. A telephoto lens brings far away objects closer to us, but it also pulls them out of their visual context. VR extends our senses by giving us a view of another world altogether; it is a controlled hallucination. Jaron Lanier has called VR "a medium that could convey dreaming" (2017, 44). Meanwhile, AR brings a part of the digital world into our physical environment, allowing us to see, hear, or manipulate digital data as if they were situated around us in our everyday world.

Following McLuhan, we can define the aesthetics of a medium as the way it conditions our senses—how we see, hear, or feel the world. The aesthetics of VR and AR depend, if only in part, on the characteristics of their underlying technology—the hardware and software. These include how 3-D computer graphics produce a peculiarly luminous, but often unnaturally clear world; how hard it is to add fine detail to this world and still redraw the graphics rapidly enough to produce a fluid experience for the viewer; the sensors that track the user's position and orientation; the limitations of the video camera on a smartphone used for AR; and so on. All these characteristics contribute to the texture of VR or AR as media and influence the kinds of experiences that can be designed. As Barba and MacIntyre (2011) have argued, the applications of AR and VR should not be understood as purely technically driven but rather as founded in a human experience of mediated space.

With VR, the emphasis is on the aesthetics of immersion and visual illusion. The makers of VR applications generally try to make their medium disappear, to make the user forget that she is wearing a headset with sophisticated hardware and software. The aesthetics of illusion is pursued through constant improvements in the technology and through design techniques that encourage the user to lose herself in the flow of the game or experience. The situation is different for AR because the user can still see and interact directly with her physical environment. So the aesthetic strategy of AR is more often to acknowledge that the experience is hybrid—part physical and part digital. The quality of hybridity makes AR experiences less immersive.

AR engages the user's senses differently than VR, and given the current state of the technology, AR addresses more of the user's senses. VR focuses predominantly on the senses of sight and hearing, but hand controllers help to make VR experiences both haptic and proprioceptive. Some VR systems are now capable of tracking your own hands even without controllers. With AR, however, the user's sense of touch is inevitably engaged along with her vision and hearing. She holds the phone and moves it around to see the digital information display on the screen. Her sense of proprioception comes into play as well when she turns to look at a digital object that may be behind her. AR is what we will call *polyaesthetic*, bringing multiple senses to most every stage of the experience (Engberg 2014).

In VR, the emphasis is on the aesthetics of immersion and visual illusion. AR is what we will call *polyaesthetic*, bringing multiple senses into play and merging the physical and virtual.

We are using *polyaesthetic* to refer to how mediation relates to your perception of the world around you and whether the media form you are using consciously calls upon one, two, or many senses. Understanding the experience of AR and VR as polyaesthetic is meant to bring experience and perception into focus, rather than technologically determined definitions of AR and VR. In an AR experience, by definition, you need to pay attention both to the physical world around you and to what is happening in the AR-mediated experience. The virtual material blends with, relates to, or in other ways asserts itself into that physical space. With reality media, the intricate relationship between human experience, mediation, and technological capabilities is always a question of balance and intention. The key question is: What is the focus of a particular experience? In a GPS-triggered audio walk, for instance, you can pay attention to the environment around you and the audio that plays without having to negotiate the application interface too much. If instead you are in a museum and the application displays images, audio, or perhaps 3-D models, the experience requires several modes of interaction—several senses to be involved as you look at the screen, perhaps manipulate the interface, listen to the sound, look at the images, and pay attention to the environment around you, as is the case with the Smithsonian National Museum of Natural History's *Skin & Bones* app (Smithsonian Institution 2014).

Creative producers in each reality medium marshal its resources in particular ways to frame experiences. If we discard the assumption that reality media are supposed to disappear and leave the viewer or user in the presence of the real, then we can better appreciate the aesthetics of each individual media artifact. We can understand the immersive aesthetics of VR as a representation in its own right, not as falling short of some abstract notion of the real. We can appreciate the polyaesthetics of AR without assuming that augmented objects are always supposed to blend seamlessly into our visual environment. We will examine further aesthetic features of VR and AR in the chapters that follow.

2   The History of Reality Media

In chapter 1, we traced the history of AR and VR back to the 1960s and Ivan Sutherland's Sword of Damocles, which made its user into an ungainly looking cyborg. We can trace the lineage much further back into the history of media. By placing AR and VR in this historical context, we can appreciate how they address the task that has characterized certain media technologies and forms for centuries. This is the task of capturing visual reality—or rather, of convincing the viewer that this particular medium achieves the goal of capturing visual reality better than any other.

Some writers like to claim that the Paleolithic cave paintings like those at Lascaux in the Dordogne region of France were the earliest instances not only of art but of VR as well—so many writers that it was almost inevitable that the *New York Times* and Samsung would make a VR tour experience for the caves (Rousselle, Shastri, and Mullin 2016). The media scholar Oliver Grau (2003) began his history of VR with Roman painting in Pompeii in the first century AD, where some of the villas' walls were preserved because they were buried in volcanic ash from the eruption of Vesuvius. We will begin our history, however, in the Renaissance because from that time on there has been a continuous tradition of applying techniques of illusion in order to construct visual reality.

Painting as a Reality Medium

It was in Italy in the fifteenth century that architects and painters developed the system of linear perspective in practice and in theory (Edgerton 2009). Their techniques drew from the science of optics that was part of the developing Scientific Revolution. Linear perspective was believed to make a painting as realistic as possible because it imitated the way rays of light reflected off objects in the world and could be focused on a surface. In other words, linear perspective was supposed to capture the way we "really" see the world.

Around 1425, the Florentine architect Filippo Brunelleschi performed a demonstration to show how linear perspective could make painting into a reality medium. Using the technique of vanishing points, and perhaps with the help of a mirror, he painted the Florence Baptistry on a small wooden panel from the perspective of someone standing at the portal of the cathedral looking toward the Baptistry. He also made a small viewing hole in the panel. He then stood at the place from which the perspective was drawn. In one hand, he held the panel in front of his face with the painting facing away; in the other, he held a small mirror. Looking through that hole, he could see the picture reflected in the mirror. When he took the mirror away, he could see the actual Baptistry. This procedure allowed him to test the painting against the reality of the building itself. By letting others look through the hole and then taking away the mirror, Brunelleschi demonstrated the efficacy of his technique (illustrated in figure 2.1).

The demonstration has sometimes been characterized as a Renaissance version of augmented reality (Levy 2012, 27). It might be better to claim it is the first example of what is called *diminished reality*, in which AR technology is used to overlay and therefore obscure something that the user would

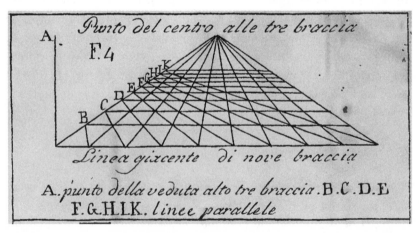

Figure 2.1
Brunelleschi's perspective technique, as documented by Leon Battista Alberti (1804).

otherwise see in the world, such as the Baptistry. In any case, Brunelleschi's configuration of the mirror and painting lacked the quality of computer-based real-time rendering and registration that are characteristic of AR. Brunelleschi had to do the alignment himself, whereas AR today performs the drawing, sensing, and tracking automatically. But the experiment was a perfect example of the La Ciotat technique: the purpose was not to forget or elide the medium of painting, but rather to confirm how well the medium constructed visual reality.

Brunelleschi's technique, based on synthetic geometry, involved drawing lines to one or more vanishing points to create a sense of perspective. Computer graphics makes this same process truly automatic, using linear algebra transformations to calculate how rays of simulated light from 3-D objects would land as points on a plane in front of the user. What the computer does algebraically in its graphical processing unit (GPU) today, Renaissance and later artists achieved geometrically, sometimes by sketching projection lines on paper or canvas. Some later artists, including Dürer and Vermeer, were known to have worked with grids or an imaging device (a *camera obscura*), and Brunelleschi himself may have used a mirror (Edgerton 2009). As Renaissance humanist Leon Battista Alberti (2005) described it in his 1435 treatise *On Painting*, the painter makes his canvas into an "open window" through which the viewer appears to see a scene on the other side. Many, perhaps most, European paintings from the Renaissance until the nineteenth century were done using perspective techniques and other methods to support this illusion, which ironically has also been called *realism* or *realistic painting* because of the conviction that this is how the eye really sees the world. The terms *illusion* and *realism* come to mean the same thing. For centuries, Alberti's window therefore helped to define visual reality for European culture.

But if most paintings in this era aimed at achieving this effect, there was a particular kind of art, called *trompe l'oeil*, that went further. The most impressive trompe l'oeil of the period were frescoes on walls or ceilings of churches or large public rooms, such as the vault of the Sant'Ignazio Church in Rome painted by Andrea Pozzo, which features a fresco that makes it look as if the vault opens to the heavens. If a viewer stands in the right spot, the perspective lines up perfectly, and she cannot tell where the physical architecture ends and the painting begins. Trompe l'oeil locates the viewer in a hybrid space that is part physical and part virtual (figure 2.2).

Figure 2.2
Fresco with trompe l'oeil. Andrea Pozzo, Sant'Ignazio Church in Rome. © 2006 by
Marie-Lan Nguyen/Wikimedia Commons. Reprinted with permission.

Baroque trompe l'oeil was, in a sense, a forerunner of AR in that it combined the physical (the architecture of the hall or the church) and the virtual
(the painting). In another sense, it was a forerunner of VR in that it blended
the painting into the whole building to create a seamless 360-degree environment. A crucial difference between this Renaissance technology and
contemporary AR or VR is, of course, that trompe l'oeil is a still image,
painted with one vantage point. Wearing a headset, today's user can move
her head or change position, and the computer will adjust the perspective
accordingly. But in Sant'Ignazio, the visitor must stand relatively close to
the favored spot. Stepping out of the magic circle where the perspective
works breaks the illusion, and the visitor is suddenly made aware of the
medium again. In this case, stepping in and out of this perspective circle is
perhaps the best way to appreciate the La Ciotat effect, a pleasant sense of
wonder at the illusion.

The Panorama

At the end of the eighteenth century, an Irishman named Robert Barker
employed perspective painting to create a related form, for which he coined

the name *panorama*. Barker's panoramas were fully immersive trompe l'oeil buildings, virtual environments consisting of a painted canvas stretched all the way around a rotunda. The viewers stood on a platform in the middle and experienced the vast painted scene in whichever direction they looked (except on the ceiling, but the roof of the building was designed to emit diffused daylight to support the illusion). In 1793, Barker built a panoramic theater in London's Leicester Square. For three shillings, Londoners could experience a 360-degree view of their own city as if they were standing at the top of St. Paul's Cathedral, which was in fact only two miles away (figures 2.3 and 2.4).

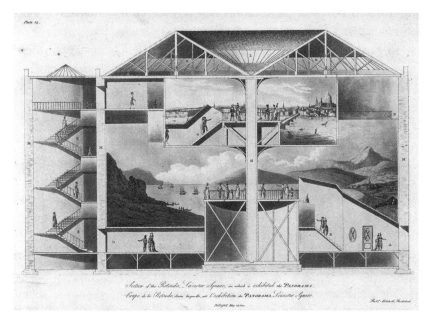

Figure 2.3
Illustration of Barker's panoramic theater (Mitchell 1801).

Figure 2.4
Robert Barker and Henry Aston Barker, *Panorama of London from the Roof of Albion Mills*, 1792.

Again, it was the La Ciotat effect, the thrill of the technology mediating reality in a new way, that attracted visitors. The thrill would be enjoyed by many thousands of viewers throughout Europe in the nineteenth century, when hundreds of temporary or quasi-permanent panoramas were erected, depicting cityscapes, landscapes, and historically important battles. In *The Panorama: History of a Mass Medium* (1997), Stephan Oettermann described how this popular entertainment phenomenon swept the continent in the century before film. Mass popularity led to many variations and developments, such as moving panoramas and dioramas, although few of these survive today.

The panoramic exhibits were not always purely virtual. In addition to the painted canvas, some exhibits had physical artifacts on the floor. A panorama of a battle might include shrubbery, rocks, and perhaps even wax figures of soldiers leading up to the painting. In Milgram and Kishino's terms, such a panoramic exhibit would be an *augmented virtuality*—closer to the virtual end of the spectrum than the physical.

Photoreality

While the panoramic exhibition is an almost forgotten reality medium of the nineteenth century, two others from that period, photography and film, continue to have a defining place in our media culture. Photography developed through mechanizing the process of linear perspective. Already in the eighteenth century, the camera obscura had been used to focus light on a surface at the back of the box or on to a mirror that reflected the image up for viewing.

The image in a camera obscura was ephemeral until the development of modern chemistry that made another medium possible. The myth of photographic reality, an early version of the La Ciotat myth, was almost immediately born. As Fox Talbot, the British pioneer of photography, put it in the title of his illustrated six-part essay *The Pencil of Nature* (1844–1846), a photograph was an image "impressed by Nature's hand" (i). Despite the fact that the camera and photographic film (like today's digital sensors) were sophisticated technologies, the photograph seemed (and still seems) to have a special claim to authenticity—which is why, for example, it is still recognized as legal evidence.

The panoramic exhibition is the forgotten reality medium of the nineteenth century, but two other technologies from that period, photography and film, continue to have a defining place in our media culture today.

For two hundred years, photography has enjoyed a greater reality status than drawing or painting. This is true even when the photograph has a narrow field of view. Employing photography to make a panorama produced a medium that strengthened its claim to reality by combining the claims of each. One early technique was simply to stitch together a series of narrower photographs (figure 2.5).

Photographers later developed systems using wide-angle, rotating, or multiple lenses to capture more, or all, of a 360-degree view on the flat surface of the film. These systems had colorful names, such as the Stereo Cyclographe, the Wonder Panoramic Camera, and the Periphote, and more recently, the Hasselblad X-Pan or the Linhof Technorama. When the panorama was just a set of conventional photographs stitched together into a long strip, then there was no consistent point of view (POV), a step back from the hand-drawn panoramas of the nineteenth century. The more advanced camera systems that maintain a single viewpoint have to distort the spherical world into a flat image through some sort of optical projection. A common projection used today is an *equirectangular projection* (figure 2.6).

For decades, panoramic photos were generally displayed in a flat format, like any other photograph. The advent of the digital medium offered new opportunities for creating and displaying such panoramas. Digital cameras—some costing less than $200—can record panoramic images more

Figure 2.5
An early flat photographic panorama of Philadelphia in 1913. Haines Photo Co., Copyright Claimant. *Panorama of Philadelphia*. Pennsylvania United States Philadelphia, ca. 1913. Photograph. https://www.loc.gov/item/2007661477/.

Figure 2.6
An equirectangular projection: the image is distorted in the vertical direction moving away from the horizontal centerline. This results in visible bulges. Père Lachaise, Paris. Photo by Maria Engberg.

easily and with fewer flaws than the earlier analog systems, and computers can display them dynamically, providing a full 360-degree experience even on a conventional screen. Viewing such images in a VR headset is even more compelling. And in addition to photographic panoramas, the computer can create and display 3-D graphic panoramas. In the 1990s, video games began to use panoramas called *skyboxes* as backdrops for shooters and role-playing games. As the player moved around and turned, she could see the sky or other backgrounds in all directions. Most players today still experience their games on conventional flat displays, although the increasingly popular VR headsets make greater immersion possible.

Digital panoramas are not limited to video games. Google Street View has already captured much of the developed world in millions of panoramic images (Wikipedia contributors 2020c). When you use Google Street View to visit some other location in the world—somewhere else in your city or somewhere on another continent—you are entering a metaverse (chapter 8). Because the VR experience of Street View is anchored to this world, not to some imaginary other storyworld or game universe, it is like Barker's panorama in Leicester Square, where visitors came to experience virtual views of the very city they were in.

As a reality medium, however, digital panoramas are still Potemkin villages compared to real-time VR. Unlike the modeling of reality through computer graphics, the panorama is a remediation of a photographic (and therefore static) representation of reality. In real-time VR, each object can move and change separately. In a virtual office composed of 3-D graphical objects, for example, you might move a chair from one end of a table to another. But in a panoramic photograph of that office, nothing can change. (So-called *lightfield photographs* are different, but they can only alter the viewer's point of view, not the objects in the photograph.) Almost all the reality media we have reviewed thus far (illusionistic painting, panoramic painting, photography, and even panoramic photography) have been static. The exception was the moving panorama, a long static image that was unrolled in front of the audience to convey a sense of motion. The first fully dynamic reality media date back to the late nineteenth century and flourished in the twentieth.

Film and Television as Reality Media

In the final decades of the nineteenth century, inventors were working on various mechanisms to lend the illusion of motion to sequences of static images, among them the zoetrope, the praxinoscope, and the phenakistoscope. In the 1890s, the Lumière brothers added another to the list, the cinématographe, which is now recognized as one of the first fully successful film cameras and projectors. It is the device that produced, for example, *The Arrival of a Train at La Ciotat Station*, described in the introduction (figure I.1).

The arrival of that train in a small town on the Côte d'Azur announced the arrival of a reality medium that depended on and at the same time enhanced the authenticity of photography. Although the early black-and-white films were overexposed and jerky (recorded at sixteen to twenty frames per second), they succeeded in reproducing a facet of reality that still photography could not. Even when viewed on large screens, however, these films could not surround and immerse the viewer, as Barker's painted panoramas had done a century earlier. Rather than striving for visual immersion, film addressed different aspects of our real-world experience: motion and time. Synchronized sound was added to film in the late 1920s and early 1930s. Although color film processes existed in the first decades of the twentieth century, most commercial movies were generally shot in black and white until the late 1930s (notable color productions included *The Wizard of Oz* and *Gone with the Wind*), and it wasn't until the 1950s that color became common. With the addition of sound and color, the conventional wisdom was that film technology had reached a certain sense of completion. (We note later in this chapter that Maxim Gorky's two complaints about the Lumière brothers' film were that they were silent and that they were in ghostly black and white.) But innovations and refinements to formats and to sound and color quality have continued.

A conventional photographic or film camera looks at the world through a single lens, but human beings have stereoscopic vision. The photographic stereoscope became popular as early as the 1850s, and already in the 1920s the film industry was experimenting with techniques for stereoscopic movies (Zone 2007). In the 1950s, these experiments resulted in a brief golden age of 3-D (Rogers 2013; Zone 2007, 2012). If you were willing to put on polarized glasses (and millions were), you could watch the scaly Gil-Man menace a beautiful female scientist in *Creature from the Black Lagoon* (1954) or the two lions lunge out of the screen in Arch Oboler's 1952 *Bwana Devil* (figure 2.7).

But the thrill of experiencing this aspect of visual reality captured in film form waxed and waned, and various 3-D or curved-screen formats with yet more colorful names (Cinerama, Space-Vision 3D, and Stereovision) failed to last. Eventually, IMAX in the mid-1980s established a niche and developed into the significant film form that it is today (Rogers 2013). Other forms of 3-D presentation requiring glasses have become almost required for animated films and some kinds of Hollywood blockbusters. These can all now be part of the experience of going to the movie theater.

A film, like a photograph, is always a recording of some past moment. Whether it is fiction or documentary, we know that whatever we see on the screen must have happened in the past. Even a sci-fi story that purports to take place on another planet in a distant future was actually filmed at some moment in our terrestrial past. Film is in this sense a nostalgic medium, always inviting the audience to look back. There has long been a desire for a medium that is truly present, one that would permit distant communication in what we now call *real time*. As the term *real time* suggests, instant

Figure 2.7
Audience members mesmerized by *Bwana Devil* on November 26, 1952, at the Paramount Theater in Hollywood. © J. R. Eyerman/The LIFE Picture Collection via Getty Images. Reprinted with permission.

distance-annihilating communication could make a medium seem more authentic, more real. And there was evidence of the desire for a real-time medium even while film technology was being developed. A well-known illustration in the British magazine *Punch* from 1878 satirically envisioned the invention of the "telephonoscope," which would transmit images as well as sound (figure 2.8).

This illustration and descriptions in science fiction predate the practical realization of television by several decades. When it did come, television was generally a one-way communication device, rather than a two-way videophone. (As it turned out, the videophone would not become viable until the arrival of Internet-based systems like Skype and FaceTime.) Crude and then increasingly refined systems for broadcast television developed in the 1920s and 1930s. The BBC began regular broadcasts in the 1930s to thousands of receivers (Wikipedia contributors 2020e); Germany offered regular broadcasts in Berlin and Hamburg (Wikipedia contributors 2020d). After World War II, in the early 1950s in the United States and somewhat later in Europe, television finally developed into a mass medium with millions

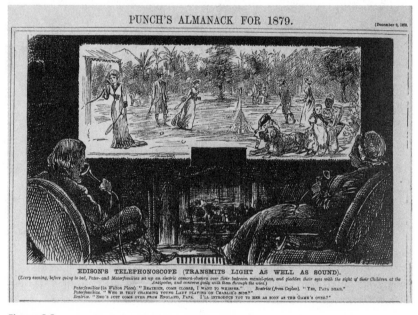

Figure 2.8
The cultural desire for television predated its invention (du Maurier 1878).

of viewers. Television had succeeded in filling the gap that film had left open for a medium that presented moving images and sound "live," more or less as they happened, even if the source of the broadcast was thousands of miles away.

The concept of liveness really only became meaningful after the development of recording technologies (Auslander 2008). Liveness is not the same thing as our lived experience of the world; it is a mediated representation of that experience. Prior to the audio record and film, concerts and plays could not be other than live. Audio recordings and films then made it possible to hear singers or watch actors when the performers were not present. Film remediated plays; audio records remediated concerts; and so liveness was born as the opposite of these new mediated experiences. But just as film and records took away the quality of liveness, television arrived to restore it. Until the late 1950s, most television was broadcast live. As video-recording technology improved, more and more primetime television was recorded, especially comedy and drama. Nevertheless, one defining quality of television was its ability to present events as they happen, and this often remains true of two characteristic television genres: the news and the coverage of sports.

It was liveness that validated television's special claim to being a reality medium. In other respects, television throughout the second half of the twentieth century still fell short of film: television was broadcast on small screens and for years in black and white, whereas the film industry developed color and widescreen formats (such as Cinemascope, Panavision, and VistaVision) in order to offer the public an audiovisual experience that was worth leaving home for. Television, of course, improved in quality too, first adding color and then introducing the VCR and the DVD, both of which further complicated the notion of liveness. The film industry has tried to keep pace. Ever since the middle of the twentieth century, the industry has continued to develop projection techniques and sound systems to offer the audience a more compelling audiovisual experience.

As digital technology began to replace analog in both film and television, these two reality media have converged—or perhaps we should say diverged—in dozens of hybrid formats. Now that almost all television shows and movies are recorded and consumed digitally, we can watch them on a variety of devices, from large LED screens at home to tablets and smartphones. Yet our media culture still refers to these digital videos as either television series or films (on Netflix, for example), based on their (presumed) original presentation as either broadcast television or in film theaters. There are now born-digital productions, such as video podcasts and YouTube channels, that seem to be both new and remediations of their two-parent reality media.

360-Degree Video

Another digital form has emerged that is in the tradition of the panorama but designed for an individual using a personal device: 360-degree video. Unlike true VR, 360-degree videos (also called VR videos or VR movies) are not generated in real time. Just as traditional movies consist of a set of photographic images shown at a rate of twenty-four or thirty frames per second, a 360-degree video consists of a set of panoramic images, each of which is an equirectangular projection (Johnson 2017; Keene 2018). Each image is displayed for only a fraction of a second and then replaced by the next one. The result is the same illusion of motion that we get when we watch a flat video on a digital screen. With the proper software, 360-degree videos can be viewed on any video screen, but they are best appreciated with a headset. All sorts of 360-degree movies have been created, many in established genres, such as documentaries, music videos, short horror films, and animated shorts, as we discuss in chapter 6.

Just as the painted panorama remediated the perspective painting and the photographic panorama remediated the photograph, 360-degree videos are clearly remediations of film and occupy an intermediate position between traditional flat film viewed on a rectangular screen in a theater and true computer graphic VR. Because they are prerecorded, 360-degree videos lack the capacity for interactivity that VR offers, but they do give the viewer greater control over her point of view. In 360-degree video, the creators lose some of the traditional filmmaking strategies or characteristics of camera angles, continuity, cutting, close-ups, and composition—the five Cs of cinematography originally identified by Joseph Mascelli in a classic work on the subject (Mascelli [1965] 1998). Some of those five Cs are still available in 360-degree films (especially continuity, cutting, and composition), but they are used in different ways than in traditional film. Directors and filmmakers have argued that 360-degree film does away with camera angles because everything is visible all the time. VR filmmaker Gabo Arora has spoken of the difficulty of learning how to edit away camera equipment and people or to hide them behind objects on the set (Sheffield Doc/Fest 2018). These manipulations, however, show that the placement of the 360-degree camera or cameras still matters, and the process of deciding where that camera will be still belongs to the filmmaker even in 360-degree video.

In the prologue of *The Five C's of Cinematography*, Mascelli ([1965] 1998) suggested a sixth C—cheating—which he defines as the artisan's skill: "Cheating is the art of rearranging people, objects or actions, during filming or editing, so that the screen effect is enhanced. Only experience will teach the cameraman or film editor *when* and *how* to cheat" (9; italics in the original). Filmmakers and cinematographers who are used to the traditional framed cinema format must learn to deal with a complete viewing space, a 360-degree round in which the viewer sees everything around her, possibly including the camera itself. There are different options for erasing the camera: someone can wear the camera on their head (figure 2.9), or the

Figure 2.9
The position of the camera in the production of 360-degree video, as opposed to VR. Here, an actress is equipped for filming *EWA, Out of Body* by Knattrup-Jensen, Damsbo, and Makropol (2019) with the camera mounted on the actor's head. Photo by Hind Bensari.

filmmaker can place the camera on a tripod and then remove the traces of the tripod in postproduction (and there are several examples in which ghostly evidence of the camera is still present).

Filmmakers making 360-degree videos still determine what the content of the whole scene is. Although they cannot control precisely where the viewer will be looking at any given moment, they can employ various techniques to try to direct the viewer's attention. They can also cut or fade from one scene to another or even move the camera during the shooting, just like in traditional film. Technological limitations also mean that the viewer of 360-degree video does not have as much visual freedom as she does in VR, where she effectively becomes the camera. In VR, both the orientation and the position of the camera are aligned with the user's phone or headset. In 360-degree video, the viewer can change her orientation, but she cannot interact with the people and objects in the video; she cannot walk among or around them. Everything in a 360-degree video remains distant from the viewer, no matter how near she might seem to be.

Reality Media as Remediations

We have now fast-forwarded through about six hundred years of reality media from the Renaissance to the present. Figure 2.10 presents a timeline for the media we have discussed.

As each new reality medium appeared, it borrowed from earlier media, while at the same time claiming to improve upon its predecessors. Renaissance painting used techniques of linear perspective to make the image realistic from a particular viewpoint. But a painted canvas on the wall was only convincing if the viewer did not look beyond the frame. The trompe l'oeil painting pushed the idea of perspective illusion further by situating the painting on a wall or ceiling to look like an extension of the architecture

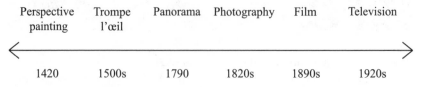

Figure 2.10
Timeline of reality media.

itself. A painted panorama was a trompe l'oeil extended to a 360-degree format. Photography remediated perspective painting in a different way by bypassing the painter, and film remediated photography by adding motion. Each reality medium added technical or technological innovations to previous forms. A linear timeline, however, cannot capture the complex remediating relationships among all these media.

As each new reality medium appeared, it borrowed from earlier media, while at the same time claiming to improve upon its predecessors.

Some reality media are static and others dynamic. Some consist of fixed images, including painting, panoramas, and photography. Some consist of moving images that are not under the viewer's control, including film, television, and 360-degree video, and also a number of "failed" media that were once popular but are now forgotten, such as the zoetrope and phenakistoscope. All these are dynamic, but not interactive. Broadcast television images can be live, but the viewer still has no control over what she will see. Like film, the images in 360-degree video are prerecorded, but the viewer has more control over the point of view. True VR belongs on the far end of the spectrum because it is potentially fully interactive.

Another distinguishing feature is whether the medium is unified and immersive or hybrid and partial. This spectrum is similar to the Milgram and Kishino spectrum, in which all mixed reality forms are hybrid and VR is unified and immersive. VR creates a complete graphic world, which is all the user can see, but in all forms of MR (mixed reality) the user sees at least part of her lived world, as well as the virtual objects or information. We can plot many other reality media on the same scale. Painting, trompe l'oeil, photography, and television are all partial. A painting in oil is a unified representational medium, but when hung on the wall of a gallery, it becomes part of a larger hybrid scene for the visitor. Painted and photographic panoramas, 360-degree video, and VR are fully immersive. Film viewed in a theater is not fully immersive, but by darkening the rest of the hall this medium effectively shuts out most of the physical world.

In one sense, every reality medium that we have been discussing is only two-dimensional. They all reduce images to a flat (or perhaps curved) screen to be viewed. The screen may be very large, as in the case of IMAX screens in theaters, or as small as a smartphone's screen or the eyepieces of a VR or AR headset. But the three dimensions are always reduced to two. The distinctions here have to do with how the images were produced and when. Two-dimensional flat reality media were made by drawing on a flat surface (painting) or reducing the three-dimensional world to a flat image photographically (in photography and film). Three-dimensional reality media use techniques of computer graphics to represent 3-D models and then reduce them. This gives the image a different texture: 3-D animation in Pixar films is easy to distinguish from the 2-D animation of decades of cartoons, from *Gertie the Dinosaur* to Disney's films of the 1960s and 1970s. And 3-D models make it possible to create perspective images in real time so that users can walk around objects in VR and even pick them up and move them. Unlike the shadows to which Gorky compared film, 3-D objects in VR can themselves cast moving shadows in their scene.

As we have noted, VR is often portrayed as the ultimate reality medium, but so were perspective painting, photography, and film in their time. It is certainly true that VR is unique in its capacity to fashion an interactive, responsive 3-D world around the user. But every one of the earlier reality media also developed a unique aesthetic, derived both from its formal qualities and its evolving place in our media culture. The differences in formal features, audience, and cultural function explain why so many of these media still survive. Of the principal reality media that we have listed, only painted panoramic exhibits are more or less obsolete. All the others still form part of today's complex media economy and are still entering into cooperative and competitive relationships with each other. Some of these media are certainly less popular than they once were. Broadcast and cable television must compete for audience and cultural status with streaming video such as YouTube and with social media in general. And all reality media have developed digital versions, which has led to hybrids. If we watch a movie on a tablet, some of the features of film as a reality medium are changed or lost. And if we watch a movie in a virtual theater while wearing a VR headset, what reality medium is that?

All remediated reality media and their hybrids redefine reality in the same way that film did when *The Arrival of a Train at La Ciotat Station* astonished the Parisian audience in 1896—that is, by asking the audience to compare the new construction of reality that they have to offer with an earlier, now familiar medium. When he viewed the Lumière brothers' films, Gorky was clearly aware of their relationship to photography—so much so that he was disappointed rather than astonished: "There are no sounds, no colors. There, everything—the earth, the trees, the people, the water, the air—is tinted in a gray monotone: in a gray sky there are gray rays of sunlight; in gray faces, gray eyes, and the leaves of the trees are gray like ashes. This is not life but the shadow of life, and this is not movement but the soundless shadow of movement" (Gorky [1896] n.d.).

Gorky understood these differences between his lived world and the world of film as failures. Film did not live up to its implied promise of reality. Other early accounts of film emphasized instead how lifelike it was, which is presumably how the La Ciotat myth gained currency, although Gunning argued that unlike Gorky, most audiences of early films understood and appreciated both the apparent realism of these films and the obvious fact that film was a medium. Over the following decades, the shadowy silence that disturbed Gorky developed into the characteristic film aesthetic of the silent period, both realistic and symbolic and abstract at the same time. We have a different understanding of that aesthetic today because of the development of film from the talkies on and because of subsequent reality media, especially television and now VR. Enthusiasts of VR today fall easily into the rhetoric of the La Ciotat myth (that VR is or can become unmediated reality), which does not allow them to fully appreciate the texture of VR as a medium. In the following chapters, we turn to the technical qualities of both VR and AR that help to define their aesthetics.

3   3-D Graphics and the Construction of Visual Reality

With both VR and AR, reality is constructed in layers consisting of a variety of media: 2-D text and graphics, photographic images, 3-D graphical objects, video and animations, and audio components (sound or music). In this chapter, we focus on the layer of 3-D computer graphics, the qualities of the space that CG creates, and the ways it remediates early visual media.

The illusion of three dimensions is key to VR and often AR, but VR and AR are not the only digital media that depend on computer graphics. Our media culture first became familiar with the shapes and textures of computer graphics through video games and special effects in films. Audiences were impressed, sometimes astonished by CGI in films starting in the late 1970s, and especially since *Terminator 2* (1991) and *Jurassic Park* (1993) in the early 1990s; video games in the 1990s were also displaying growing sophistication in creating graphical worlds. In film, the goal is often to make the CGI effects blend seamlessly with the live-action photography so that the audience has difficulty seeing where the physical world ends and the CGI begins. In the Jurassic Park films (five to date and no doubt counting), although the audience knows that the dinosaurs must be either computer generated or animatronic (animated physical models), it wants to believe that this is what a T. rex or velociraptor would really look and act like if genetic engineering could in fact bring them back to life. The original *Jurassic Park* features a La Ciotat moment. When the paleontologist who visits the island first sees a brachiosaur that the entrepreneur and his scientists have created, he becomes dizzy and has to sit down. His astonishment at the real dinosaurs within the film stands in for the audience's astonishment at Spielberg's CGI illusion. The 1993 audience reacted to CGI just as the audience in the Grand Café reacted to the early black-and-white films, and for the same reason: both technologies seemed to go further in imitating the real than any other contemporary media. Video games too, especially first-person shooters and role-playing games, often aim for realism or hyperrealism. The creators of AAA sports games, in which the player directs the actions of a football or baseball team, strive to make the graphics look as much like a television broadcast as possible. Sometimes if you walk by someone playing an Electronic Arts (EA) NFL game and glance at the screen, you may briefly (but only briefly) be fooled into thinking it is indeed Monday Night Football.

Until the nineteenth century, the task of constructing a visual image required a human hand (chapter 2). But when Niépce, Fox Talbot, Daguerre, and others figured out how to focus light on a plate or film coated with light-sensitive chemicals, they seemed to have found a way to make images *naturally*. This conception of photography continues to influence our thinking, even in the era of Adobe Photoshop. Although we know that photos can be enhanced or radically changed through postprocessing, we feel the urge to believe in the truth of a photo unless we have good reason to think otherwise. We believe that a photo on the front page of the *New York Times* or another reputable newspaper shows us what *really* happened—but we doubt the photo of a crashed alien spaceship in a supermarket tabloid. The historical importance of photography as a reality medium made it the standard of remediation for film at the end of the nineteenth century and then again for 3-D computer graphics in the second half of the twentieth. Hence, photorealism has been the stated ambition of computer graphics experts since the field's earliest days in the 1960s and 1970s. From then to the present day, we can think of the development of computer graphics as akin to a cartoon trying to become a photograph or an animated cartoon trying to become a live-action film. This is not the same as making computer graphics into unmediated reality. It is the attempt to redefine reality with reference to another medium, just as film was thought to define reality as photography in motion.

Computing Photoreality

The first step in achieving photorealism in computer graphics lies in reproducing the perspective in which we see the world. When we look at the world, objects that are far away appear smaller than ones that are close, and straight lines converge in a particular way. Ancient Greek and Roman painters had experimented with perspective, but as we noted in chapter 2, it was Renaissance painters beginning with Brunelleschi in the fifteenth century who developed systematic techniques for drawing in linear perspective to create the illusion of three dimensions on their flat canvases. The computer can in fact draw objects in perspective with more precision than a painter can by hand, at least up to the limits of the number of pixels on the screen. The Renaissance technique was at best only approximate, and painters sometimes distorted the perspective intentionally—for example, to emphasize one part of the scene over another. Computer graphics has another advantage. By redrawing the scene repeatedly (in real time or beforehand) with progressive variation, the computer can create the illusion of motion in video games (in real time) and in animated films (beforehand). Animators working by hand could never create complicated scenes in linear perspective because it would take too much effort to draw all the frames. A ninety-minute animated film might have over a hundred thousand separate frames. They could never change the perspective from frame to frame, which could require redoing the angle of every object in the previous frame. This is the equivalent in a live-action film of panning or dollying the camera, and such camera movements are easy to imitate in a 3-D CGI animated film. The spiraling camera in the ballroom scene of *Beauty and the Beast* in 1991 (a film that combined traditional hand-drawn animation with CGI) was the first well-known example of what the computer could achieve in this area.

Computer graphics uses the remediating term *camera*, but a CG camera is a mathematical construct rather than a physical mechanism. It is simply the point that represents the viewer's eye, at a location in front of the projection plane. The computer draws the objects on the plane so that they look "right" for a viewer at that point. As the viewer, you do not have to be located precisely at that one point. If your eye is relatively close to the right position, you can adjust to see things in perspective. The same is true when you watch a traditional movie in a theater. Even if you have to sit very close to the screen or off to one side, you soon adjust so that the image makes sense.

As reality media, VR and AR decompose visual experience into a series of discrete elements and discrete steps. Everything you see has to be calculated. In analog photography and film, each still or film frame consisted of black-and-white or color areas. In CG, the scene is built of discrete objects, each composed of geometric shapes, usually triangles, that are fitted together into more complex shapes called *polygons*. Any actual scene will have hundreds or thousands of them. In a computer game like *Half-Life: Alyx*, each of the characters, the objects, and the rooms and buildings is decomposable into at least hundreds of triangles. But rendering the right objects in perspective is not enough in itself. To make the scene more convincing also requires calculating the effect of light on each object. Just as there is no physical camera in CG, there is no physical light illuminating the scene. CG manipulates lighting, remediating the effects of light in photographs and what Renaissance and Baroque artists did in their paintings. The difference is that the manipulations are by algorithm and of pixels rather than continuous tones. The lighting of the scene is simulated by changing the color of objects (pixel by pixel), color changes called *shading*. The effects of various kinds of light—ambient light, spotlights, directional lights, and point lights—must be calculated, as well as the way that the various surfaces in the CG scene reflect or refract. The rendering time for all this lighting is often an issue. In a Disney-Pixar film, the subtlety of how objects are shaded comes at the price of significant computational requirements. The individual frames cannot be created in real time. Unlike in animated films, VR and AR do have to be rendered in real time as the user watches, which is why the scenes cannot be as complex and the shading and lighting must be simplified. As technology improves, more visual effects become possible in real-time CG. As of 2020, the ray-tracing technique that leverages deep learning and a high-end graphics processor card was able to achieve sophisticated lighting and shading effects in real time (Huang 2020). Due to the massive power requirements of such graphics cards, however, this level of detail will not be available in stand-alone VR or AR headsets any time soon.

Another aspect of our world that must be created digitally is its so-called physics. When objects move in a CG animation, their motion must appear to be consistent with the envisioned world's laws of physics. Because the objects have no real mass or dimensions, their physics too must be simulated by an algorithm. Otherwise, the player of *Half-Life: Alyx* could accidently run through a wall and emerge on the other side. The effects of collisions must be simulated, as well the effects of acceleration and gravity—that is, if the animators want the scene to look and act like our world. But the physics of computer games do not have to be the same as those of the physical universe. Gravity could have any value the animators want, or it could be absent. A ball could gain velocity when it ricochets off a wall, rather than slowing down. Each effect will have its own visual meaning, creating a sense of realism or surrealism. In the era of cel animation (when each frame was drawn by hand), cartoon characters walked with a peculiarly exaggerated motion, which came to be an expected part of their visual behavior. The exaggeration, anticipation, and distortion of their movements were not true to life, but they did seem true to the cartoon world. CG animation has remediated some of those effects and invented others.

All these techniques (projection, shading, lighting, motion and physics) illustrate the same key feature of VR and AR as reality media: they construct their layers of visual reality through computation. The earlier reality media of photography, film, and television used chemical, mechanical, and electronic means to record the world and play it back to their viewers. They were all fundamentally dependent on the light of the physical world for the scenes they presented. AR still depends on the world of light as a backdrop for its CG effects. But AR and, to an even greater extent, VR construct a digital simulation of the visual world, which they process and represent to us as viewers. CG draws on and remediates elements from painting, photography, live-action film, and cel animation. The peculiarities of the way CG algorithms must digitally reconstitute all these elements of the visual world are what give CG an aesthetics that are both related to and distinct from these earlier forms. In figure 3.1, for example, the artist is drawing from conventions of nature photography, with the chameleon and branch in focus and the background blurred. The chameleon itself, however, is a fantasy creature, as so often in 3-D computer graphics, from a metallic world with a nature different from ours. The precision of CG imagery fashions a space of peculiar clarity and potentially distances the viewer from the image. Empathy does not belong to this aesthetic.

Its complex heritage pulls computer graphics in different directions. The playful media forms of animated cartoons and some genres of video games incline computer graphics toward simplified iconic figures, landscapes, and imaginary worlds, where characters and objects morph into fantastic shapes and defy gravity—a cartoon world, which is technically easier for computer graphics to illustrate. Yet, as we have noted, there has always been a strong impulse toward photorealism, toward imitating what can be achieved in photography and live-action film.

Figure 3.1
Zaki Abdelmounim, *Curiosity*, 2014. Reprinted with permission.

Photorealism and the Uncanny

In *The Matrix*, Neo inhabits a digital simulation without knowing it. He believes he is living with millions of fellow human beings in a large American city, but the city is in fact a simulation by a vast computer network being transmitted directly into Neo's brain, as well as the brains of all the other humans, who live in nutrient pods constructed and tended by machines. The simulation is visually perfect (and also perfect for the other senses), so Neo has never suspected the truth. Within the movie, the city looks perfectly real—because the scenes were shot in live action in a real urban environment (mostly Sydney, Australia). What Neo and we as viewers accept as reality at the beginning of the movie is reality as constructed by the conventions of Hollywood cinema. Ironically, when Neo takes the red pill and joins Morpheus in the "desert of the real"—the wasted ruin of a world that is actual reality in the context of the Matrix films—the actors are performing in front of green screens. This "real" world depends heavily on CG, while the illusion created by the Matrix is filmed in live action.

The term *photorealism* has been used for decades to designate visual experience as it is captured in static form by the photographic camera and in moving form by the film camera. Ever since Ivan Sutherland began to experiment with CG in the 1960s, one of the key goals of the work has been photorealism. Each advance in projection, lighting, and shading was measured by how much closer it brought CG to making an image indistinguishable from a photograph, an animation indistinguishable from live-action film. And close is not good enough. The CG world (gamers and critics, as well as designers and computer researchers) speak of the *uncanny valley*—a term coined in 1970 by the roboticist Masahiro Mori specifically to describe robots that look or act in ways similar to but not indistinguishable from humans (Mori 2012; Reynolds 2015). Mori suggested that close similarity can make robots seem eerie and therefore more threatening. Hence the valley, the drop in the affinity we feel to robots if they look almost but not exactly human (figure 3.2).

As with robots, figures in CG may look uncanny if they resemble humans, but not perfectly. And even as their visual appearance improves in

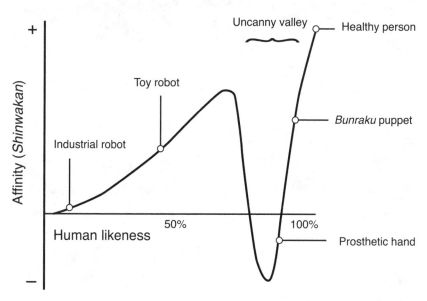

Figure 3.2
Masahiro Mori's uncanny valley for robotic devices. M. Mori, K. F. MacDorman, and N. Kageki, "The Uncanny Valley [From the Field]," in *IEEE Robotics & Automation Magazine* 19, no. 2 (June 2012): 98–100, https://doi.org/10.1109/MRA.2012.2192811. Reprinted with permission.

impressive ways (figure 3.3), animation and behavior can contribute to the uncanny (Schwind, Wolf, and Henze 2018).

The lesson commonly drawn from this eerie effect is not to remain on the far side of the uncanny valley, but to try to jump over it into complete photographic and behavioral realism. What is overlooked is that photorealism has never been the same thing as our lived visual reality. It is a measure of the real based on the characteristics and aesthetics of earlier media,

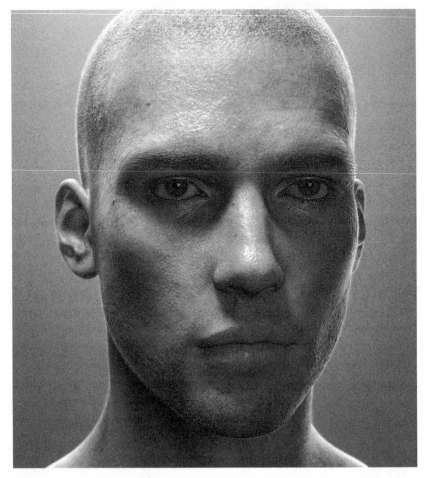

Figure 3.3
Visual realism is increasing, but the uncanny valley can remain if the subtle human animation and human behaviors are not perfect. "Mr Head" by Chris Jones. https://www.chrisj.com.au. Reprinted with permission.

photography, live-action film, and, to some extent, television. This is in the nature of remediation: a newer medium depending for its definition of reality on a familiar and successful older one. For example, EA's football, baseball, and basketball games try to look like a televised broadcast, so they appropriate many of the camera techniques as well as other conventions of that medium. EA sports games remediate television viewing as much as the experience of being in the stadium. Indeed, if an EA game perfectly imitated what it was like to be sitting in the stadium watching a baseball game, you would be stuck in one seat, unable to see the action from various angles and parts of the stadium or to replay a home run. Such a video game would never sell. Instead, the camerawork and the announcers calling each play bring the game closer to its goal of remediation.

Photorealism is not the same thing as visual reality. It is a measure of the real based on the medium of photography. This is how remediation works: a newer medium depends on an older, familiar medium for its definition of reality.

Where video games have been using computer graphics on a player's monitor to borrow from and refashion television and film, VR now suggests the tantalizing possibility of offering the player a fully immersive photorealism. With their emphasis on interactivity, VR games are, above all, remediations of the desktop or console video game as a medium. The conventions of plot and action, however, are often drawn from film and television, even if traditional camerawork is no longer possible. In games and other VR applications, photorealism is one of the techniques for fostering a sense of presence, a feeling of "being there" in a fully CG world (see chapter 5).

Beyond Photorealism

Although photorealism has been the holy grail of computer graphics since its inception, this is not to say that all computer graphics aim exclusively at photorealism. One of the vaunted appeals of VR is that it can free developers to reimagine the world. The VR pioneer Jaron Lanier likes to say that in VR you can become anything you want—a dragon, for example, or a teapot. One of his definitions of VR is "a medium that could convey dreaming" (2017, 44). In AR experiences as well, designers may sometimes try to blend computer graphics seamlessly into the real world, or they may allow those graphics to stand out and become visually distinct. In both VR and AR, computer graphics has the potential to cover a large range of representational styles—as large, for example, as hand-drawn comics.

In *Understanding Comics* (1993), Scott McCloud, both a practicing cartoonist and a theorist of comics, provides a taxonomy for comics that is also useful for computer graphics. He depicts the range of possible styles as a triangle, the three vertices of which are realism, abstraction, and symbolism (or, as he labels them in figure 3.4, reality, the picture plane, and meaning).

Comic artists have been successful across much of this range—from the soap-opera style of a Mary Worth to the highly iconic Mickey Mouse. As McCloud points out, they have tended to avoid the extremes of realism

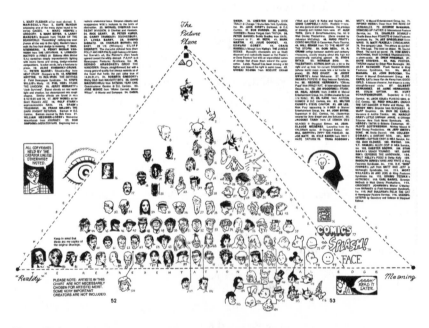

Figure 3.4
Scott McCloud's picture plane. Reprinted with the permission of HarperPerennial, *Understanding Comics: The Invisible Art* by Scott McCloud. © 1993 by Scott McCloud.

(left) and simple line drawing (right), and especially abstraction (top). Comics do often include text, purely symbolic representation, the lower-right vertex. For the medium of hand-drawn comics, the reality vertex would not in any case be photographic realism, but rather the tradition of realistic European drawing and painting. McCloud's triangle could in fact be applied to other visual media, such as painting. Until the middle of the nineteenth century, most "serious" painters kept relatively close to the reality vertex of McCloud's triangle. With the birth of modernism, painting in general moved away from realism and up toward the abstract vertex. Today, visual artists feel free to explore any point at all within the triangle—from realism to abstraction to verbal symbolism.

The same is true of 3-D graphics, in which artists and producers may work anywhere in the triangle. CGI in mainstream films is usually blended seamlessly with the live-action photography, as in *The Matrix*. The representational and aesthetic strategies of video games, whether played on a monitor or with a VR headset, are more varied. Action-adventure games like *Half-Life: Alyx* tend to stay in the lower left of McCloud's triangle, with avatars and storyworlds that look photorealistic—or rather, hyperrealistic, as they are often set in postapocalyptic wastelands and other planets or dimensions. Other genres, such as platform games and abstract puzzle games, are more stylized or cartoon-like. As for VR and AR, although their graphics can occupy any place in the triangle, virtual experiences tend to keep toward the bottom. AR may move toward the right and be more iconic. Both AR and VR make use of text to apply labels to objects or places in the physical environment. Such labels reflect the fact that AR and VR can draw objects in both two and three dimensions. Flat text, images, and even videos find their way into 3-D graphic space. The coexistence of 2-D and 3-D objects in VR and AR worlds reminds us that computer graphics is ultimately a mathematically generated illusion, sometimes unlike anything we experience in the world of light.

As we have suggested, it is the aspiration of many, perhaps most, CG researchers to push 3-D graphics as far as possible into the realism corner of McCloud's triangle. As we also noted, a fully convincing CG replica of our visual world will not be possible on VR headsets in the foreseeable future. To put on a headset is to enter into a world characterized by the special aesthetics of computer graphics.

In the decades since the heroic age of the 1970s, the applications of computer graphics have grown increasingly diverse. There are now numerous commercial and scientific uses, many of which may not require photorealism—for example, medical imaging and data visualization, computer-aided design (CAD), and prototyping in industry. Even entertainment does not always aim at photorealism. Whole genres of video games stay firmly in the middle of McCloud's pyramid, the realm of visual fantasy. In the endless Marvel Comics or Transformers movie franchises, the producers do not want the effects to look completely natural. A giant metal robot would be stiff and, well, robotic, in real life. So-called nonphotorealistic CG effects have been available for decades in 2-D graphics in applications such as Adobe Photoshop. Many of the thousands of Adobe Photoshop filters are designed to take a photographic image further away from photorealism—to make the image look like a watercolor painting or pencil sketch, to give it a sepia tone or high contrast. Similar effects can apply in 3-D as well.

Hybrid Representations

The Marvel Comics and Transformers films remind us that computer graphics can be used in hybrid representations that do not aim at seamless integration. Most AR applications fall into this category. It is in some ways even harder to achieve a photorealistic blending of computer graphics and the physical environment in AR than it is to build a world from scratch in VR—and most AR experiences do not even try but instead present us with two clearly distinguishable layers of the virtual and physical. When *Minecraft Earth* displays building platforms in a yard or a park, no player really confuses the Minecraft objects with their real-world settings. Part of the charm of the game is that these stylized objects appear to occupy space in the world in a way that even the youngest players know is impossible.

Movies have been preparing our media culture for such hybrids for decades. In the early days of Hollywood computer graphics, *Who Framed Roger Rabbit* (1988) combined live action and traditional animation with a fluidity that audiences had not seen before. The movie was a live-action comedy, but the premise was that the cartoon characters or toons featured in animated shorts lived in the same physical world as humans in the city. The toons bounced convincingly off physical walls and ceilings, while a human detective rode through real streets in a cartoon roadster. Just as the visual aesthetics was a combination of live-action and animation, the plot combined outlandish cartoon comedy with film noir. The lead toon, Roger Rabbit, is framed for a murder, and the human detective investigates and clears him. In this film, CG special effects helped to make the integration of the cartoon characters with the actors so effective.

The tradition of layering animation into live action dates even further back—to the period not long after live-action film was invented. From 1911 on, the animator Winsor McCay introduced a film-going public to the idea of animation in a series of films, including the 1914 *Gertie the Dinosaur*. This was a live-action film featuring McCay and his friends and illustrating the genesis of the animated Gertie. The audience got to see the dinosaur herself move and to see McCay interact with his drawing (figure 3.5).

Through the decades that followed, film has offered combinations of these two very different visual regimes: such as the stop-action animation that allowed King Kong to scale the Empire State Building with Fay Wray in hand in 1933 and, decades later, Jason to fight the Hydra in *Jason and the Argonauts* (1963). These were in turn predecessors of the animatronic and CG dinosaurs in *Jurassic Park* (1993). Various techniques have played on

Figure 3.5
Winsor McCay introduces Gertie the Dinosaur.

the audience's desire for photorealism. They worked precisely because audiences wanted to regard film as the record of something taking place in our world. A Hollywood movie is a fictional story, but its actors are real people, delivering their lines before a camera at a particular moment in a particular location. Film experiences like *Gertie, Jurassic Park*, and the contemporary superhero franchises locate us in a reality somewhere between live action and animation. They fulfill a fantasy that we have as children, in which what we can imagine becomes visible and operative in the physical world. They occupy the reality corner of the pyramid and somewhere in the iconic middle at the same time.

Few AR experiences (so far) are as imaginative as *Jurassic Park*, but AR does function in the same way between two representational regimes. This double mode is not limited to AR games; even the IKEA app that allows us to visualize how furniture would look in our living room transforms that room into a hybrid space. An app might display the names of all the restaurants or hotels around us, putting each name above or in the direction of the appropriate building. Google Street View now offers an AR feature, Live View, that combines text and icons to help the user navigate.

Reality Media and Uncanny Doubling

Doubling is an essential feature of AR and, in another sense, of VR as well. Every computer-generated world in VR is a stand-in for the physical world that we have left behind when we put on a headset. Which brings us back to the uncanny valley and the quest for photorealism. As we noted earlier, the standard view in computer graphics is that the uncanny valley is a place users do not want to visit. They want to remain on the side of cartoon-like graphics or eventually to jump over the uncanny valley into perfect photorealism, which would mean making characters and landscapes that are indistinguishable doubles of the real world. The examples that we see in figures 3.3 and 3.6 are eerie because they are not on either side.

Unwanted and dangerous doubles have been a mainstay of horror stories for centuries—tales of the undead, such as vampires and zombies, or creatures made by human hands, such as the golem of Jewish folklore, and endowed with something that approximates life. In the nineteenth century, obsessed scientists and technicians started making such doubles, in literature such as Mary Shelley's *Frankenstein* (1823) and E. T. A. Hoffmann's

"The Sandman" (1816). It was The Sandman that Freud invoked in his early twentieth-century essay on the uncanny, and such uncanny doubles remain common today in TV series such as *The Walking Dead* (2010–present) and films such as Jordan Peele's *Us* (2019).

The reaction to the uncanny, however, can be something other than horror. The desire to achieve photorealism in CG and human realism in robots can easily become sexualized. A science fiction film entitled *b*, reported to be in development, will star an android actress named Erica, created by roboticist Hiroshi Ishiguro to be "the most beautiful woman in the world" (Bahr 2020). Erica looks in fact like an expensive sex doll, an uncanny (and unsuitably racialized) object of desire (figure 3.6).

And although the technology of robotics and machine learning that animate Erica is cutting-edge, the theme of desire for an uncanny sexual partner is not new. In E. T. A. Hoffman's nineteenth-century short story "The Sandman," the young hero Nathanael is tricked into falling in love with

Figure 3.6
Sam Khoze and Erica. Photographed by Elizabeth Sadegh. © LIFE Productions Inc. Reprinted with permission.

the automaton Olimpia and goes insane when he discovers that she is not really human. As for contemporary computer graphics, 3-D CG pornography is becoming a not insignificant part of the genre, along with live-action 360-degree video (see chapter 6). A fascination with uncanny doubles does not, of course, have to go as far as sexual desire. Such figures or robots are uncanny because they are familiar and unfamiliar at the same time. We could almost believe that they are photographs and not CG models. Their subtle flaws fascinate us, and that fascination is a version of what we have been calling the La Ciotat effect.

Such a reaction is not limited to photorealistic effects. In playful ways, *Gertie the Dinosaur* and *Who Framed Roger Rabbit* also feel uncanny because cel animation or CGI can convey a sense of motion and apparent life. A sense of the uncanny always depends at least in part on the current media technology. The audience in 1896 was certainly more likely to be amazed by a black-and-white film of a train than we are today because of our familiarity with color films and wide-screen formats. Thus, in twenty years, when VR headsets no longer evoke a strong sense of amazement, a subtle feeling of the uncanny can remain.

That feeling is essential to all reality media. Uncanny doubling is implied in the phrase "reality media" itself because these media always double reality in some way. VR and AR are uncanny, and so are film, television, and photography, each measured by the standards of their contemporary media world. Attempts in certain genres of VR games to fashion perfect photorealistic doubles of the world through computer graphics do not elide the medium, but instead make us all the more aware of VR as a medium and the process of doubling. Every VR experience is uncanny, and not just because the representation falls short of the goal of being a perfect copy. As we noted in chapter 1, Lanier claims that VR experiences give him a new appreciation of the "nuances" he "used to be blind to in the light and motion of the natural world" (2017, 285). Every VR experience reflects some aspect of our natural world, even if it is a shooter game that takes place on another planet. And there is also always a gap, a difference between the representations in VR and the world as we experience it when we take the helmet off.

The uncanny is essential to all reality media. Uncanny doubling is implied in the phrase because these media always double reality in

some way. VR and AR are uncanny, and so are film, television, and photography, each measured by the standards of their contemporary media world.

The *awareness of the uncanny* is thus another name for the La Ciotat effect, which is rooted in a conscious or unconscious acknowledgment of the difference between each medium and other media. We have described in this chapter several ways in which the use of CG can be deployed to remediate earlier media forms. In both desktop and VR video games, CG can create a visual space, the uncanny clarity of which contrasts with the various styles of live-action film. CG techniques can combine the capacity of live-action film to show a world that looks uncannily familiar with the exhilarating possibility of cartoon animation that any imaginable world can be created and visited. At the same time, the CG camera remediates and ostensibly improves on the physical film camera by giving the viewer freedom to explore that world. For AR, CG can integrate the digital data of the internet into the visible space all around us: the immersive web liberates the World Wide Web that was previously trapped on our screens.

CG in film, VR, and AR create visual aesthetics that differ from earlier forms and so represents reality in a new way, just as the earlier reality media of television, live-action film, and photography did when they were new. But there is no escape from the uncanny valley. That is not how reality media work.

4 Degrees of Freedom: Spatial Tracking and Sensing

Chapter 3 was about the space that computer graphics constructs; this one is about finding the user of an AR or VR system in space. The two questions here are: Where is the user in the world (spatial tracking), and what is the world like around the user (spatial sensing)? The first is the key to making an AR or VR experience responsive. The user can effectively examine and move through the space of the experience only if the computer can keep track of where she is and where she is looking. The second question is the key to a new set of AR (and potentially VR) applications. Improved cameras and software for computer vision are making it possible for mobile devices to "understand" the world that the cameras see. Applications begin to see *into* the world, marking surfaces such as floors and walls in three dimensions and eventually being able to distinguish furniture, humans, and all the other objects the room contains.

Tracking is important for both VR and AR, but in different ways. In VR, because CG constitutes the world for the user, a VR system needs to know where the user is and how she is oriented relative to that graphical world. But to do that, the system needs to track users in the physical world locally— perhaps just their orientation or their position in a very small area. In AR, however, where the graphics typically form only a sparse layer overlaid on the user's visual world, the system must keep virtual objects positioned (or *registered*) relative to places and things in the physical world. The problem of accurate registration has been a principal research area in AR for decades and has held back the widespread use of AR as a medium for many kinds of applications. We have entered a new era in AR applications precisely because tracking and sensing technologies are becoming more accurate and less expensive. With the introduction of Apple's ARKit in 2017 and Google's ARCore in 2018, AR on mobile phones has been able to achieve the stable placement of digital objects in space. AR headsets such as Microsoft's Holo-Lens 2 can do the same.

These tracking and sensing systems are finally starting to work in the way that people have always assumed AR does work. As we have said, most people have gotten their notions of both augmented and virtual reality from decades of depictions on film and television. VR has been more popular, but AR has also been featured in films from *The Terminator* (1984) to *Minority Report* (2002) to *Iron Man* (2008) and many others. In such films, the AR is flawless: digital objects are integrated perfectly into the space of the live-action characters. What can be achieved in the movies through hours or days of rendering of CG special effects is only now beginning to become possible on smartphones and tablets in the hands of real users. With stable tracking, augmented reality has arrived as a medium, responsive and available to millions.

The new tracking and sensing techniques work in the way that people have always assumed AR should work based on film and television.

TV and AR

Combining text, graphics, and live action in the televised image became common decades ago, although the practice was generally limited to chyrons, program titles and credits, and advertising. Then in 1992, CNN began dividing the screen into windows—the news anchor in one window, a reporter in another, and statistics or crawling text at the bottom. This hybrid scheme was probably influenced by the multiple windows of information in the graphical user interfaces of the Macintosh computer and Microsoft Windows operating system in the 1980s and 1990s. But in general, each window on the television screen still contained only one kind of representation—a text, an image, or a video.

An important moment in the integration of graphics and live action came in the fall of 1998, when ESPN started displaying a digital first-down line in its broadcasts of football games. This was perhaps the first widely seen and successful example of AR on American television, and the "first-and-ten" system is still used. Viewers know that the yellow first-down line, the blue line of scrimmage, and other graphics are not really painted on the football field, although few are likely to reflect on how the blending works. A computer does the blending in (near) real time as it processes the camera feeds before sending them out to the world.

The technique has migrated to other sports. Visual overlays in sports car races identify each car, its driver, and its speed as the cars round the track. In baseball, the strike zone and location of the ball are overlaid after each pitch. In addition, virtual advertisements are projected on the fences behind the plate. These compositing techniques are not full interactive AR. What viewers see on their television sets or phones is still just a video stream; they cannot enter into the space and move around the virtual images. But such broadcasts have accustomed a vast audience to the layered aesthetic that is the essence of AR.

Tracking the User

The purpose of tracking in AR and VR is to adjust the visual scene in accordance with the user's movements. As we described in chapter 3, in computer graphics, there is an algorithmic camera that serves as the viewpoint for the object's scene. Each time the user moves in an AR or VR application, she is in effect changing the position or direction of the CG camera. If the user is wearing a headset, then she moves the camera by turning her head or by walking around. If she is looking through the screen of the phone, she moves the camera with hand gestures. In either case, the system must track her movements and quickly (ideally thirty or sixty times a second) redraw the CG objects in the scene accordingly. The viewer has two kinds of visual freedom: in orientation and in position. She has three *degrees of freedom* in orientation (along the three axes of pitch, yaw, and roll, shown in figure 4.1a). She can also keep her orientation steady and move her whole body: forward or back, to the right or left, or up and down. She thus has three degrees of freedom in spatial positioning (figure 4.1b).

To account fully for her movements in space, an AR or VR system needs to track all six degrees of freedom. This tracking constitutes what is called the viewer's *pose*. In VR, the pose determines where the user appears to be standing and looking at the CG objects in the virtual world. In AR, the pose is where the user is in fact located and where she is looking in the physical world around her. Knowing her pose enables an AR system to integrate a layer of virtual objects into her view of the physical world.

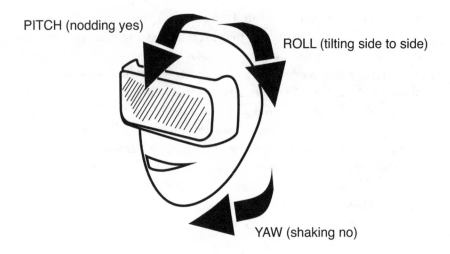

3 degrees of freedom can be tracked by the goggles.

(a)

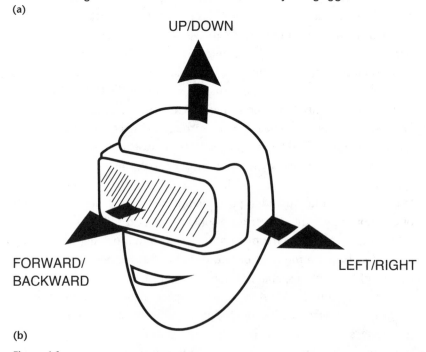

(b)

Figure 4.1
(a) The viewer's orientation; (b) the viewer's positioning. Graphics by Carina Ström Hylén.

In both VR and AR, pose is essential to the medium's construction of reality, although the effect is experienced differently in the two media. In VR, the freedom of point of view seems more magical, perhaps because it enables the viewer or player to travel freely through the space of an often-fantastic VR world. In AR, the world that the user sees is the world of lived experience, in which she is accustomed to being able to move freely. To give the user this liberation, the system must surveil her in a way and to a degree unlike any previous media. When the tracking functions properly, it is seamless and does not attract the user's attention per se; rather, it is the manifestation of seamless tracking that the user experiences. In VR, as noted, this is the freedom of movement in the virtual space. In AR, it is instead an absence of movement—the fact that the virtual objects stay fixed in her visual space even when she turns her head. The digital portal depicted in chapter 1 remains in the same apparent place in the street as the user approaches it and passes through. These effects typically demand precise registration of objects against the user's visual background, which in turn requires techniques of sensing and of tracking. VR generally only needs to know where the user is, whereas AR often needs to know more—not only where the user is but also where the world is and what it looks like.

The kind and amount of world information that an AR system needs depends on the application. When we use phones to provide directions as we drive, the application needs to know our absolute position in the world. It can rely on GPS to determine where the phone is on the surface of the earth and compare these data with stored road maps. The phone is equipped with a GPS chip to receive the signals, perhaps assisted by triangulation from cell phone towers or local Wi-Fi, and the positioning only needs to be accurate to a radius of many feet. (To display these directions using AR, much more precise and stable positioning is needed, beyond what GPS is capable of.) To play a tabletop AR game or to pop a 3-D model onto an advertisement in a magazine, the application needs the position of the phone relative to the table or the magazine. It does not need to know the absolute position of the table or magazine in the world. Techniques for either absolute or relative positioning or both are needed for all AR applications; for precise positioning on phones and usually on AR headsets, one or more cameras (and other sensors) need to be gathering visual data from the world. As the techniques are refined, the phone or headset learns more and more about the world through its sensors.

Google, for example, is developing a system that achieves precise location tracking outdoors by enhancing the GPS location with data from its vast repository of Street View panoramas. Its Visual Positioning System (VPS) uses the phone's camera to scan the area around the user for identifiable features and compare them with Street View images to determine where the phone is with significantly greater accuracy than GPS alone (Reinhardt 2019; TechFunnel 2020). This is a technique for so-called global localization in that Street View is extensive, so VPS should work for a large portion of the developed world, especially in urban environments. Global localization will be a key component for constructing AR mirror worlds in the future (chapter 7).

Sensing the World

One technique for learning about the world, image tracking, has already been in use for years. If the phone's camera sees something it can recognize (say, a photo in a magazine ad), then it can make that image an anchor point and draw virtual objects in relation to the anchor. As the user moves the phone, the anchor point will move in the camera's field of view, and the system can adjust the virtual scene accordingly (figure 4.2).

AR systems are increasingly capable of operating anywhere without relying solely on a database of prestored images by finding their own anchor points in the visual environments. Software and hardware working together are getting better at techniques such as simultaneous localization and mapping (SLAM), using features they find in the world to build a map of the world around the user and then track the user's motion relative to the map. Useful in a number of high-tech applications, including self-propelled robots and self-driving cars, these techniques are the foundation of modern AR platforms. Beyond just tracking, these techniques can understand more about the world. The simplest of these is called *surface detection*. Flat surfaces such as floors, desktops, and walls are detected automatically and can be used as anchors. As noted, the anchoring techniques, combined with so-called visual inertial odometry (VIO), is what enables the portal discussed in chapter 1 to remain stable on the street. Surface detection is a powerful technology in itself, enabling all sorts of AR applications. For example, games such as *Minecraft Earth* use surface detection along with

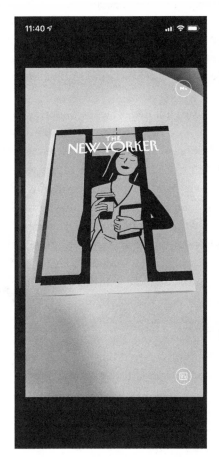

Figure 4.2
Image recognition: The cover of the *New Yorker's* May 16, 2016, issue interacting with an AR application. The camera finds the photo through the app and displays the 3-D animation. © Christoph Niemann. Reprinted with permission.

absolute positioning through GPS to anchor its build platforms in the player's yard or in a park.

Surface detection is only the first stage in the more ambitious idea of building a 3-D model of the visual world (see chapter 7). As SLAM and other techniques mature, AR systems can detect more about the environment around you. Instead of just flat surfaces, the 3-D structure of the world can be recreated. Eventually, simple models of the things can be recovered, and these real-world models can be treated in the same way as other CG objects so that virtual objects and physical ones become part of a unified scene. In an outdoor AR game, a virtual character could be partly hidden behind a physical wall. Virtual signage could be obscured behind a physical building until the user turns the corner.

The mapping process can progress toward full-fledged scene understanding. Rather than just generically recognizing a table, the systems will eventually be able to identify what kind of table, and so on for all the objects it sees—cars, buildings, people. The system would know which things were movable and which were fixed, which were animate and which not. A wide variety of applications—not just games, but other commercial and scientific applications as well—become possible. An AR application with semantic scene understanding could help a surgeon anticipate and identify problems during an operation (Pauly et al. 2015), in which the patient's body, as well as the surgeon's tools, constitute the scene. A similar application could help a mechanic diagnose and repair an airplane engine. Some imagine that a user could hold her phone up as she walked around her apartment, and the application would recognize the layout of each room, the furniture, and the objects and store the dimensions and locations of everything. It could send all this information up to a server owned by one of the giant social media or search companies, which would then be able to correlate this information with user profiles, friends, and online shopping habits. This scenario suggests a new dimension in service, as well as a potentially massive threat to security and privacy, as we'll discuss in chapter 9.

POV and the Aesthetics of AR and VR

As we noted earlier, a key feature of VR and AR is the way tracking and sensing enable them to redefine the user's point of view. Both VR and AR take control of the camera out of the hands of the producers and give it to the viewer, at least to some degree, and this ceding of control is one of the salient differences between these two new reality media and the earlier film and television. The experience of watching a film in a darkened theater may be engaging intellectually and emotionally, but it is physically passive. Decades ago, film theorists compared the situation of the film viewer to Plato's allegory of the cave (Baudry 1975). The darkened theater is like Plato's cave, and the film audience members are figurative if not literal captives, in the sense that there is nothing of interest to see except what the flickering cone of light shows on the screen in front of them. If in watching film or television the viewer can see only what the director decides to show her, in VR and AR, she becomes the camera so that the very notion of POV changes. While a film director typically chooses to mix in an occasional first-person POV shot with establishing shots, over-the-shoulder shots, close-ups, and so on, the whole experience in AR and VR is (usually) one continuous POV shot.

This is one reason that tracking is so important to these new reality media: the user can only become the camera if the system knows her pose. If the pose were not connected to the camera, the user in VR would not feel any sense of immersion in a virtual world, and the user in AR would not feel that the digital information were connected to her physical world. Film and television function independent of the viewer's point of view. In a movie theater, the projection technology cannot determine and adjust to where each audience member is sitting, so it has to project the same images in the same perspective for everyone. Its reality is fixed long before the audience enters the theater. With a live television broadcast, the image is created in near real time, but again without the user; the image is the same whether the viewer is looking at the screen or away from it.

In film and television, the viewer's gaze must move as the camera moves, but with VR and AR it is the reverse. Film and television control the viewer's perceived motion through all six degrees of freedom, and the viewer is usually seated and does not move much on her own as she watches. The user of a VR or AR experience may also be seated. But even then she is not chained and compelled to look straight ahead, like Plato's prisoners. She moves her phone or her head to exercise her three orientational degrees of freedom, and she has controls (buttons on the keyboard or hand controllers) to move or fly to change position. In many applications, she can physically walk within a limited space in VR or anywhere she likes in AR. Even now when watching a film or TV show, viewers may experience the motion of the camera proprioceptively, in their bodies. Watching a filmed ride on a rollercoaster can make them queasy, especially in a more three-dimensional format such as IMAX. The effect is stronger in VR. When the user's vestibular sense fails to confirm the motion that her visual system is reporting, the contradiction makes her nauseated. This simulation sickness (see chapter 5) arises because VR puts users in an uncanny valley of proprioception. VR engages their bodies to a far greater extent than television, although not as completely or naturally as when they engage with the physical world.

In VR, our pose connects our physical bodies to a digital, graphic world. In AR, our pose connects both our own bodies and the physical world around us to the digital. VR is often said to be the quintessential immersive medium. It is not only 360-degree computer graphics that convey a sense of immersion, but also the sensing and tracking that permit the user to explore a virtual world. In AR, when technologies such as VIO and SLAM make possible the combination of computer graphics with the physical world, they have the effect of incorporating the physical world into a new hybrid. Even without AR, our smartphones have been turning the world into a hybrid medium for years: every time we look through the screen to take a photo or a video, we see a media object, the world as photo or video. An AR application fashions a world in which digital artifacts (texts, 3-D graphics, or audio) blend into, complement, and reinterpret the meaning of the physical. Two worlds (physical and digital) that had existed in separate spheres come together through sensing, tracking, and ultimately scene understanding.

Tracking and sensing are thus integral to the aesthetics and to the functioning of AR and VR, integral to defining the experiences that the media offer. They extend our senses by sensing us in return, our position and movements in space. They fashion a space in which digital materials (text, 2-D and 3-D graphic objects, and video and other media) become dynamic and responsive to the user. This responsiveness contributes to the La Ciotat effect of these new media. In VR, tracking allows us to explore the computer graphic space in ways that make VR worlds seem both alien and familiar, natural and artificial. In AR, both tracking and sensing enable media artifacts (especially information taken from the internet) to become a part of our lived world while remaining within the digital realm at the same time.

5 Presence

You put on an HTC Vive headset and other equipment. You are led to a physical black pillar in which indentations simulate the curves of a hole in a giant sequoia tree. This is Treehugger: Wawona, *an interactive VR installation created by London-based creative studio Marshmallow Laser Feast. What you see in VR space is the giant tree: "As you touch the trunk of the enormous tree, you can feel the edges of a knot in the bark. If you lean your head into it, your vision passes through the surface of the virtual sequoia. Inside, you can see the tree's xylem pump water from the ground up through its trunk. Then you begin to levitate, following the flow of water towards the canopy. Looking down is disorientating, so you cling to the physical surface of the tree" (McMullan 2016).*

Treehugger: Wawona (figure 5.1) from 2016 is a perceptually rich experience that transports you to the forests of giant redwood and sequoia trees on the West Coast of the United States. It was featured at the Tribeca Film Festival in 2017 and won the Tribeca Storyscapes Award. Along with 360-degree video, VR or interactive film exhibitions have become an integral part of festivals such as Tribeca and Sundance. *Treehugger: Wawona* has been described as strange, psychedelic, and disorientating (McMullan 2016). It does not rely on photorealism, and yet it gives you a sense of intimacy with the tree—with its bark, with the cells that make up its being. The imagery was partly captured in forests through *photogrammetry*, a technique by which computer vision is used to construct 3-D objects from photographs. The makers call it a *digital fossil*: a representation of the life of the tree gathered from various sensing techniques, including LIDAR and CRT scans. The result becomes something other, something more uncanny than a superficial 3-D

Figure 5.1
Treehugger: Wawona—experiencing the life cycle of a tree in VR. © 2016 by Marshmallow Laser Feast. Reprinted with permission.

model of a tree. It is a digital double constituted from various kinds of data that in this form allow the audience to touch, feel, smell, and sense the life force of a giant sequoia tree.

The vividness of the work illustrates what VR experts have called *presence*. But it would be naive to think that we experience the sequoia in *Treehugger: Wawona* as if we were in the presence of the real tree. High-resolution graphics and responsive tracking can contribute to a sense of immediacy, but even the best VR technology cannot deceive users into believing that they are having an unmediated experience. There is another way to think of presence. Instead of an *as if* feeling, it is a feeling of *both and*; that is, the experience is *both* mediated *and* immediate at the same time. We never entirely forget that we are having a VR experience, but we find ourselves on the threshold of forgetting. Being on that threshold is an uncanny feeling, a sense of presence in a reality medium.

> **VR and AR cannot deceive their users into believing that they are having a nonmediated experience. But that is not necessary for a sense of presence.**

Definitions of Presence

In the 1990s, VR developed its own research community in computer science, and from the beginning, presence and immersion became the measures and the elusive goals of VR. There were no smartphones that could be used to create mobile VR experiences. The user would have to wear a headset attached to a computer or enter a VR cave. In either case, the virtual environment would surround and envelop her. The question was whether she would believe in the CG world that she saw. Would she forget that she was actually sitting or standing in a laboratory? Would sensory immersion lead to a feeling of presence?

In the first issue of the journal *Presence* in 1992, one of the founding editors defined *presence* as the feeling of "being there" (Sheridan 1992). Many other definitions have been offered since (Cummings and Bailenson 2016). Already in 1997, two researchers, Lombard and Ditton, offered a classification of different ways in which VR can condition how we perceive and experience the world, and their classification is still useful today. They grouped definitions into two broad categories: (1) individual perception of the world and (2) social interaction and engagement with others.

The first category includes presence as transportation, as immersion, and as realism. VR is not the first medium to transport us. Many of the reality media that we surveyed in chapter 2 make use of their affordances to transport the viewer somewhere else. As soon it became a mass medium in the 1950s, broadcast television, for example, was celebrated for its capacity to take us to the places where news and events were happening. The pioneer television journalist Edward R. Murrow opened the first broadcast of *See It Now* in 1951 by transporting viewers in real time from the Brooklyn Bridge in New York to the San Francisco-Oakland Bay Bridge in California. A few years later, Walter Cronkite hosted *You Are There*, which transported viewers back in time and space with pseudo news reporting of events like the sinking of the Titanic or the siege of the Alamo. VR today makes its special claim to transportation through encompassing and dynamic representations of *there*: its 360-degree CG imagery together with user interaction. The Windows Mixed Reality system launches by transporting you to its "homebase": a mountaintop villa, which is a portal through which you can experience a metaverse of VR games and interactions. The key to this form of presence is to suggest that your body has left its digitally mediated cocoon and traveled to a villa or, as in *Treehugger: Wawona*, to the range of sequoia trees in California.

A VR head-mounted display with headphones and handheld controllers creates an environment that is meant to fully occupy your attention. There is nothing that you see other than what the computer draws. If the headset includes earphones, the system's audio mutes the outside world more or less completely. Your sense of touch and proprioceptive embodiment are not completely engaged, but the controllers do draw you further into the virtual space. VR hardware and software are constantly being improved to intensify the sense of perceptual immersion. But again, VR is not the first medium to aim for this effect. Watching a film in a darkened theater is an immersive experience, even if the immersion is not as complete as with VR. IMAX or 3-D film systems get closer to visual immersion. There are some 4-D theaters or theme park rides that build physical effects to complement what the audience sees on the screen—for example, synchronized motion seats and environmental effects such as water, wind, fog, scent, and snow (4DX 2020).

From a historical perspective, the most salient measure of presence is the degree to which a medium can produce realistic representations of objects and events. To say that a reality medium achieves presence by being realistic seems like a hopelessly circular definition. But the claim makes sense once we unpack how the word *realistic* is being used: a reality medium achieves presence by comparison—by convincing its user that it is better than any other available medium or technique for representing reality. In chapter 2, we surveyed how realism was pursued through this comparative strategy for perspective painting, panoramic exhibitions, and film. It is what the Renaissance Leon Battista Alberti was doing for perspective painting when he claimed that this technique can make a painting into a window onto a world. Similarly, the word *photorealism* makes photography the standard of realistic representation today. We saw in chapter 3 how computer graphics appeal to the standard of photorealism. As VR and AR gain more importance as media, CG itself might one day become the standard. As early as the 1970s, computer graphics pioneer Alvy Ray Smith was said to have defined reality as eighty million polygons per second (Rheingold 1991). Before the development of CG, no one would have thought to define visual reality in terms of polygons.

There is another dimension to realism beyond accurate visual representation, which has to do with our understanding of the world we live in. In the Jurassic Park movies, for example, the dinosaurs look uncannily realistic. But our sense of what is possible in our world assures us that these dinosaurs could not have been on the set with the live actors. (No velociraptor wranglers were needed.) They had to be computer-generated and composited into the scene. Even a high degree of visual realism will not convince us that we are in the presence of the real. What *Jurassic Park* and its sequels achieve instead is a sense of delight at how this visual technology can create the illusion of the real.

These qualities (transportation, immersion, and realism) all center on our individual perception. A feeling of presence can also arise when the medium provides a connection with other people, even without high-quality sensory immersion. This link between presence and social engagement is certainly not limited to 3-D VR; many contemporary media can be used to generate a sense of connection. A phone video of Hong Kong protesters being attacked by the police might arouse a feeling of identification and presence, if we imagine what it would be like to be in that crowd under attack. A closely related notion is what Lombard and Ditton called *social richness*, the awareness of being one actor among many in a social environment. Multiperson VR environments for entertainment and virtual conferencing (such as Mozilla Hubs and AltspaceVR) are becoming popular, and such VR systems offer an experience different from video conferencing. The participants in a VR environment are represented by avatars that occupy the same 3-D space and can provide some social, symbolic, and nonverbal cues lacking in video conferencing. Sharing virtual space in this way may produce a (modest) sense of the presence that would be felt by the people who are conversing in the same physical room.

These definitions circle around one core idea: that presence is a kind of absence, the absence of mediation. Presence as transportation, immersion, or realism all come down to the user's forgetting that the medium is there. If the medium can effectively disappear, then the user can be transported to another world. If the representation is realistic, she will feel a part of that world. If she is fully engaged with other human actors in Rec Room or Hubs or even with so-called nonplayer characters in a video game, she may temporarily ignore the headset or the screen she is using. The idea is that if the user can be enticed into behaving as if she were not aware of all the complex technology, then she feels presence. Forgetting the medium remains the goal for many enthusiastic developers of 360-degree video and true VR experiences.

Presence and Empathy

In 2015, two filmmakers, Chris Milk and Gabo Arora, created a 360-degree video called *Clouds over Sidra* (Arora and Milk 2015), produced with the support of the United Nations in an effort to raise awareness of the Syrian refugee crisis by showing the life of refugees in the Za'atari camp in Jordan. The camera follows twelve-year-old Sidra in her everyday life, bringing the viewer along into her school, a gymnasium, a bakery, and the tent that is now her family's home. The goal of this kind of 360-degree film is, in the words of Arora (Le Forum des images 2017), "not storytelling; it is storyliving." He claimed that photography does not provide a strong enough connection to the children and their plight to elicit compassion and ultimately action. Photography and film just do not work anymore, Arora suggested, because in a media-saturated world, we are long-since numb to these kinds of visuals. Milk made a similar claim in a TED talk (Milk 2015), in which he called VR an *empathy machine*, appropriating Roger Ebert's term that originally described cinema. Milk said of *Sidra*, the movie: "When you're inside of the headset . . . you see full 360 degrees, in all directions. And when you're sitting there in her room, watching [the girl Sidra], you're not watching it through a television screen, you're not watching it through a window, you're sitting there with her. When you look down, you're sitting on the same ground that she's sitting on. And because of that, you feel her humanity in a deeper way. You empathize with her in a deeper way."

Milk makes a typical claim of remediation when he compares 360-degree video to television. Television is a mere medium, but 360-degree video in a headset, Milk suggests, allows you to be present with Sidra (Engberg and Bolter 2020). For Milk, 360-degree video brings you closer to reality; its greater authenticity is demonstrated by the deeper empathy that you feel. Lisa Nakamura (2020) has argued that Milk's hyperbolic statements are not only misleading but damaging. Instead, she suggests that these refugee documentaries in 360-degree format create a "toxic empathy that enables white viewers to feel that they have experienced authentic empathy" (47) and that rather than gaining an understanding and learning by listening to the experiences of the refugees themselves (in these documentaries, often women or children of color), 360-degree documentaries such as *Clouds over Sidra* risk "confus[ing] immersive viewing with access to the actual experience" (54).

A claim similar to Milk's has been made for immersive VR, in which researchers have long pursued the idea of emotional reactions as a test of presence. For example, researchers have designed VR experiments to induce in subjects the same anxiety toward heights that they would experience in the real world. A subject is put in a virtual world in which she must stand at the edge of a virtual pit and look down. If her heart rate increases, this is taken as evidence that she feels a sense of presence (Meehan et al. 2005). There have also been studies regarding VR and empathy. Subjects who were put in the virtual perspective of the homeless showed greater empathy toward their plight (Herrera et al. 2018). The VR artist Nonny de la Peña has created several similar experiences—not as scientific experiments, but as political statements on imprisonment in Guantanamo and hunger in Los Angeles (see chapter 6).

In all these cases, VR is understood as getting us closer to the authentic or the real by surpassing other media in achieving presence. But as we suggested at the beginning of this chapter, forgetting the medium is not necessary for a sense of presence, which can instead be understood in a more nuanced way as a liminal zone between forgetting and acknowledging VR as a medium. Understanding VR as a reality medium requires an account of how these VR experiences are representational and remediate earlier genres and conventions, at times with problematic effects.

AR and Presence

What about presence in AR? Returning to the first definition, the feeling of *being there*, we could say that AR always generates presence, because the user is never cut off from the environment as she is in VR. If you are experiencing AR on your smartphone, the physical world is there for you to see just beyond the screen. When you wear a headset such as HoloLens 2, you may feel more enclosed, but you still look through the eyepieces at the surrounding world. Your sense of being in the world is hardly compromised.

Presence also figures differently in different kinds of AR experiences, as the following examples illustrate.

> *You are reading one of the* New York Times*'s immersive articles with augmented reality, which, when read on a mobile device, offers a mixed reality experience. Digital 3-D graphics material is presented alongside text. You're reading an article about an exhibition of David Bowie's costumes (Ryzik 2018). A 3-D model of David Bowie's 1973 lightning bolt suit (figure 5.2) is displayed on a table in the café in which you are sitting.*

This experience can be had anywhere, as long as there is a large flat surface upon which to anchor the costumes. If you peek around the screen, the costumes disappear, but through the screen, there is a high degree of realism. This illustrates presence as realism, though not as immersion or transportation. You are not transported anywhere. If anything, it is the reverse: the costumes are transported from the cloud to the café.

> *You are in Athens, Greece, walking through the agora, the ancient marketplace that was the hub of activity, where people came to meet and shop. Today it is in ruins, with the foundations of buildings from which stubs of columns rise. But with the aid of a tablet and AR software, you can point your camera at the remnants of a building called the Middle Stoa, and a portico with Doric columns will appear in the place where it stood in 180 BC. You can examine it from a variety of distances and angles against the background of the contemporary ruins (Verykokou, Ioannidis, and Kontogianni 2014).*

In this experience, the sense of presence comes from knowing that you are on ground the ancient Greeks walked. Your sense of presence evokes both the here and now and the past at the same time. The feeling of being in a special place is a feature of many AR experiences that contrasts with all VR. Presence here also derives from the realism of the 3-D modeling of the ancient building, which is joined relatively seamlessly to that ground.

> *You are a mechanic about to perform maintenance on a BMW Mini R56. You launch the HoloGARAGE application in your HoloLens 2 and let it scan the car with the hood open to register the device's position relative to*

Figure 5.2
A 3-D model of David Bowie's 1973 suit displayed through the *New York Times* app.
© 2018 New York Times.

the engine and carriage. Then you see a series of menu items floating in the air. One choice allows you to access the manual, another the wiring harness, and so on. You choose the Bolt and Torque item. When you look at the car's wheel, you see the proper torque to apply on your wrench to loosen each of the bolts. The HoloLens 2 continues to provide overlay information as you maintain the brakes (Yamagata Corp. 2017).

None of Lombard and Ditton's definitions of presence apply particularly well to the car repair example, except perhaps a notion of psychological immersion. If the system is working properly, the mechanic can focus on the task of checking the brakes, not on the AR interface. The interface becomes transparent, and the mechanic feels fully present in the work.

You are visiting New York City, and you are using Google Maps to navigate your way around Manhattan. You have directions for the Museum of Modern Art and can see the route on your screen. But you're not entirely sure whether you are walking in the right direction. So you switch to the AR mode, Live View, which allows you to hold the phone up and see arrows that indicate the way to go. As you walk, the arrows update to keep you on the right path.

This is AR as an annotation of the world around you, one of the principal services that an AR mirror world (see chapter 7) may provide. The presence you feel in this case is not transportation to another place but verification of where you are. The arrows are not meant to blend into the scene. The whole point of this kind of application is that the signs and annotations will stand out from the streetscape in front of you.

Place and Space

The preceding four examples each represent a larger class of AR experiences. Some AR apps can work anywhere. The *New York Times* app will situate David Bowie's costumes (Ryzik 2018) or Winter Olympics stars (Branch 2018) on any surface that the reader chooses. AR tabletop games just need a flat surface, or in some cases a special marker on that surface, to situate the game. However, you can only see the Stoa rise from the ruins if you are standing in one place in Athens, Greece. The HoloGARAGE app should recognize a BMW Mini in any garage in the world. The Google Maps Live View will work wherever Street View has adequate data, but it gives directions specifically based on where you are.

Two of these classes of experiences are specific to places in the world (such as Athens or New York City), and two are simply responsive to the space around them (Bowie's costumes and HoloGARAGE). The distinction between meaningful places and arbitrary spaces in the world was articulated as early as the 1970s by the geographer Yi-Fu Tuan ([1977] 2001). Places are spaces that have special personal or cultural significance for us, and geolocated AR can mobilize that significance for us. Two of the four classes are experiential, meant to provide compelling experiences. It may be cool to stand in front of David Bowie's costumes; it may be exciting to stand before a 3-D replica of an ancient Greek building. The other two are practical. You may be eager to get to the Museum of Modern Art, but the Google Maps Live View is not in itself a compelling experience. The same is true of the HoloGARAGE app: it is designed to help a mechanic get a job done. We could plot these and other AR experiences on a 2-D graph, with place and space on the horizontal axis and experience and information on the vertical (Rouse et al. 2015).

These four cases are designed for a single user. Like VR, AR has the potential for social presence as well, although this potential is only beginning to be developed. For many AR applications, such as games, the other participants may also be physically present in the same space. Children play *Minecraft Earth* with friends around them working on the same virtual structure. AR technology provides the opportunity for social presence. Microsoft's Holoportation system is specifically designed for telepresence: two people in different locations seeing 3-D dynamic images of each other as if transported into their own space. Each avatar is a photorealistic 3-D image of the other person, not a Pixar-like avatar (Microsoft, n.d.). This is presence through (reverse) transportation and realism, but, above all, through social interaction—bringing another person into your social space.

AR applications supporting multiple users and offering social presence could occupy any of the four quadrants of our graph (see figure 5.3). Two or more mechanics could work together on the same car; a whole tour group could visit the Athenian agora together; and so on.

VR applications can be situated on the same graph. VR games, such as *Star Wars Battlefront Rogue One* (Electronic Arts), can be highly engaging

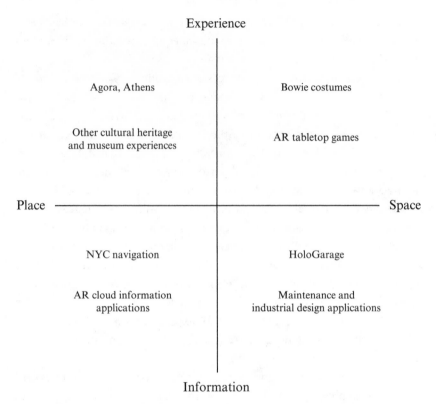

Figure 5.3
Different classes of AR applications.

experiences. Systems like Hubs or AltspaceVR are also generally experiential rather than informative, although you can, of course, engage in a serious conversation in them or even give a lecture. During the COVID-19 pandemic, VR conferencing became increasingly important as an alternative to physical travel. VR simulations and training applications fall on the informative side of the graph. Most VR applications belong in the right half, taking place in an anonymous, CG space or, in the case of some VR games, a fictional space of science fiction or dystopian fantasy. However, there are VR simulations of real places. Tourist or educational VR applications could take visitors to the Acropolis or the Inca ruins at Machu Picchu. In that sense, VR can occupy all the quadrants of the graph. Still, there does seem to be an important difference between a simulation of the Athenian agora in CG and actually standing in the agora seeing an AR representation of the Middle Stoa. That difference is captured in the term *aura*.

Aura

This heightened form of presence was described by Walter Benjamin long before the advent of computers. In the mid-1930s, Benjamin wrote "The Work of Art in the Age of Mechanical Reproduction," in which he introduced the concept of *aura*. He started from the fact that most works of art throughout history have been unique. An artist produces a painting by hand, and each painting is one of a kind. The same is true for a sculptor working in stone. In each case, the product comes into being at a moment in time and in one place. Aura is the *here and now* that each such work possesses because of its unique history of production and transmission. It names the sense of authenticity and originality we attach to such works of art. A desire for aura is the reason that millions of people still visit the Louvre each year and strain to catch a glimpse of da Vinci's *Mona Lisa* (behind glass) when they could buy a high-quality reproduction of the painting to hang in their own living rooms. Even if your copy were a commissioned version in oil by a skilled painter, it would lack the *aura* of the original.

When Benjamin was framing the concept of aura, there were already two important "new media" that were different: photography and film. (Television also existed in the 1930s but was not yet widely available.) He argued that, unlike painting or sculpture, photography and film do not inspire feelings of reverence and remoteness because they reproduce scenes and objects "automatically" through the use of a camera. Furthermore, individual photographs and films can themselves be reproduced automatically in an arbitrary number of (nearly) identical copies. For the viewer of Charlie Chaplin's *Modern Times* (1936), the experience is the same no matter which of the copies of the film she happens to be viewing. In the age of film and photography, Benjamin contended, aura undergoes a cultural decline.

We can bring Benjamin's concept into the present and ask about the status of VR and AR. It seems clear that VR generally lacks aura. Most VR experiences are in theory completely repeatable wherever the VR equipment can be set up. In a pure VR application, the actual physical location of the user is usually irrelevant. Ideally, the user neither sees nor hears anything in the room where her body is situated, experiencing instead a wholly virtual world.

AR is more complicated (Bolter et al. 2006). Because AR applications are not purely virtual, they are not perfect *reproductive* technologies. Some draw on the physical and cultural uniqueness of particular places, but not all AR has aura in this sense. For a tabletop AR game, it should not matter what room you are in; you just need a flat surface and enough space. Even when you are using the Google Maps Live View to navigate through the city, you are not experiencing aura from the app itself. When AR is used for visiting a museum, a historic home, or an archaeological site, however, it is taking advantage of the aura of those places. An AR app may even enhance their aura in the sense that it could make the user more aware of the cultural significance of a picture on the wall of the museum or the historical significance of a battlefield or public space.

When AR is used for visiting a museum, a historic home, or an archaeological site, it is taking advantage of the aura of those places.

Cybersickness and the Negation of Presence

We noted earlier that VR researchers have sometimes read physiological responses as an indicator that a person is feeling present in a VR experience (Meehan et al. 2005). For many users, VR, and to a lesser degree AR, can cause physical discomfort, no matter the level of sophistication in visual representation or technical perfection. The term *cybersickness* has been coined to describe symptoms including headache, nausea, eyestrain, dizziness, fatigue, or even vomiting that may occur during or after exposure to a virtual environment (Rebenitsch and Owen 2016). Unlike motion sickness, cybersickness does not require actual movement and was first recorded in flight training simulators (Davis, Nesbitt, and Nalivaiko 2014). What causes the sickness is not yet fully understood. The most popular scientific explanation is that it is due to a mismatch in sensory input. Two key sensory input systems are engaged in virtual environment experiences: the visual and vestibular senses. They both give cues to position, orientation, and perceived movement, and what the individual sees in a virtual environment does not correlate with what she feels is happening physically. If she sees a virtual view that suggests that she is riding on a rollercoaster, but her body is not moving, this causes a sensory conflict, the theory goes, to which the body reacts. There is a wide variety of research on cybersickness, examining different types of moving images and various displays, considering factors of age, gender, and physical ableness, and so on (Arafat, Ferdous, and Quarles 2016; Stanney, Fidopiastis, and Foster 2020). Other studies have suggested that rather than sensory mismatch, it is a question of postural instability (Stoffregen and Smart 1998). This theory argues that humans attempt to achieve postural stability and that virtual environments disrupt this control.

While the underlying physiological mechanisms that cause cybersickness may still not be fully understood, many of the factors that are part of why users develop symptoms are. These include factors related to the individual user, such as age, gender, physical abilities, posture, and temporary illnesses such as the flu; device factors that link the technology to issues such as lag, flicker, field of view, calibration, and ergonomic features; and, finally, task factors, which include the duration of the experience or the level of control that the user has (Davis, Nesbitt, and Nalivaiko 2014; Narciso et al. 2019; Rebenitsch and Owen 2016.

The type of imagery matters as well. The illusion of self-motion without actual movement, or *vection*, increases while watching moving scenes or scene oscillations, and this in turn is known to produce sickness. The key to solving the issue of cybersickness in VR lies not only with the technical systems, Davis, Nesbitt, and Nalivaiko suggest, but in the ways in which virtual environments promise "a natural way of interacting with computers using the human body and all its senses" (2014, 1). Experiences in VR ask the user to suspend belief in her lived reality in favor of a virtual one. The various embodied sensations and discomforts that follow for many VR users suggest that an intricate balance of belief/disbelief is not fully achieved. Even with habituation, an embodied uncanny feeling may linger.

Cybersickness is visceral evidence that VR is not the medium to end all media. It is naive to believe that VR finally achieves what all earlier reality media were purportedly striving for: perfect transparency. By combining 3-D graphics with sound and eventually haptics, VR may achieve a degree of transparency; it may astonish us. But that astonishment is circumscribed both by our bodily reactions and by our awareness of other media. Cybersickness reminds the susceptible user of the medium in a powerful way. Nausea replaces astonishment. Even if improved hardware and software can ultimately diminish or even solve the problem of cybersickness, VR will still be experienced in the context of our complex media culture. We will take off the VR headset at some point and return to that complex culture, in which VR has to compete with other media. We will evaluate its claim to presence in relation to those other media, and this suggests a more nuanced definition of presence. It is not forgetting the medium completely; rather, presence is a sense of wonder that comes through a subtle balance between forgetting the medium and acknowledging it. In that sense, remediation is as much a part of presence as sensory immersion.

Thinking back to our first example in this chapter, *Treehugger: Wawona*, we might ask what it remediates. The very fact that it was exhibited at Tribeca and other film festivals tells us what the producers think. This is a VR experience that offers a challenge, or at the very least a complement, to more traditional filmmaking as a way of experiencing the world. The producers are asking us to see *Treehugger: Wawona* as a new kind of documentary, a kind of science film in which the viewer/user enters into and learns about the tree in a new way. Some of their later work, such as *We Live in an Ocean of Air*, suggests that the human viewer's position is in fact decentered in favor of nature and animals instead: "Using a unique combination of technologies from untethered virtual reality, heart rate monitors and breath sensors to body tracking, visitors will be completely immersed in a world beyond human perception" (Marshmallow Laser Feast 2018). The presence these works offer is to transport us beyond the surface of the object, to become part of the tree in a way that film cannot, while still reminding us of our own bodily presence.

6   The Genres of AR and VR

This chapter is a survey of current AR and VR applications. Given the rapid development of VR and AR, many of the examples we list are bound to have been surpassed by others by the time you read this. Nevertheless, the genres that we describe (games, art, visualization, navigation, and so on) have already been around for several years, some in the laboratory if not in general use, and are likely to remain and grow for the immediate future. Most current AR and VR applications behave like applications or artifacts that we know from earlier media. *Minecraft Earth* was unmistakably a video game. *Clouds over Sidra* is clearly a documentary. Reality media applications often function as additions to established genres. This is most obviously the case for the categories of entertainment and the arts. Video games have been played for decades on consoles and computers. Documentaries in 360-degree form clearly belong to a film tradition that began in the 1920s and 1930s. Both VR and AR contribute new forms to the genre of media-supported conferencing and chatting that dates back not only to telephone and video conferencing for business but also to the party-line telephone systems of the early 1900s. Even the scientific and commercial uses of AR and VR are often remediations and belong to genres. An app that lets you visualize a piece of furniture in your living room is an extension of the product catalog or in-store furniture displays. An AR headset that permits building inspectors or first responders to see through the walls of a building is a remediation of paper blueprints and wiring diagrams.

All these applications are media, whether for entertainment or work, and they remediate in the sense that their viewers or users understand them as belonging to a genre of applications that deliver a certain kind of experience or serve a particular task. Within their genres, AR and VR draw on earlier media both for content and techniques but also seek to offer a more compelling experience than their predecessors. They forge new conventions made possible through the medium's characteristics. We see this pattern of borrowing and creative refashioning repeated in many application areas, including games, artistic expression and storytelling, documentary and journalism, tourism and cultural heritage, visualization for medicine and architecture, marketing, social media, and pornography.

This chapter, the longest in this book, still cannot adequately cover all the types of existing or planned applications. The applications that we include are not the only ones that exist today, or even necessarily the most popular or successful. They are examples of the genres that we have observed, and they suggest emerging services; experiences; opportunities for the digital mediation of entertainment, pleasure, work, and learning.

Video Games

In the summer of 2016, a revival of a popular 1990s video and trading card game, Pokémon, led to a Pokémon Go craze throughout the world. This augmented reality game for mobile phones captured the attention of news media trying to fill thin midsummer pages. In July 2016, then thirty-three-year-old father of four Sam Clark, from Southampton, allegedly became the first in the United Kingdom to catch the 142 virtual creatures that were placed all around the UK (Taylor-Kroll 2016). He noted that in the process of trekking around in search for the virtual creatures that were geolocated around his city, he lost weight and found new friends in fellow players.

Your first encounter with both VR and AR is likely to be through games. Perhaps you have tried games themselves, like the AR game *Pokémon Go* or the VR games that are featured now for all the major systems (Windows Mixed Reality, SteamVR, Oculus, Sony PlayStation VR). Perhaps you have read about such games or seen them represented in films or advertised on TV or YouTube. VR and AR games have the potential to attract many of the hundreds of millions who play other forms of games, which is why forecasters count games as the most important growth area for both VR and AR in the coming years.

Pokémon Go is perhaps still the best-known AR game, in large part thanks to the media coverage of the *Pokémon Go* frenzy of the summer of 2016.

When you visit the *Pokémon Go* locations (scattered all over the world), you see wild Pokémon on the screen of your phone as if they were physically in the world, and you capture a Pokémon by flinging a virtual ball at it. Downloaded hundreds of millions of times, it was the application that defined AR for a broad public (Barrett 2018; Robinson 2016). The success and fame of the game was probably due in no small part to the existing Pokémon brand. There had been other, earlier AR games. In fact, Niantic, the company that produced *Pokémon Go*, modeled the game on its earlier AR game, *Ingress*, which had met with limited success in comparison (Sabin 2017). The *Pokémon* franchise was already decades old, and this was certainly part of the answer for the AR game's surprising impact. It was the first *Pokémon* game on a mobile phone and the first free *Pokémon* game on any platform. In any case, *Pokémon Go* successfully exploited the classic features of AR, allowing players to interact with digital objects located in the world.

When VR and AR games remediate earlier games, what they add is a feeling of immersive presence, in the case of VR, and often physical presence, in the case of AR (chapter 5). What they borrow from earlier games may include existing storylines, graphics, audio, and modes of interaction that can serve as models or be directly repurposed. And video games, like any other media form, come in a wealth of genres and subgenres; Wikipedia lists dozens (Wikipedia contributors 2020f). VR and AR games reflect and remediate this multitude. Even at this relatively early stage of adoption, the variety of available titles is impressive, especially for VR (Moore 2020).

As we would expect, there are already many VR shooter games and action-adventure games (which often amount to some navigating and problem-solving between episodes of shooting). From the 1990s, many gamers have favored first-person shooters and role-playing games with increasingly elaborate 3-D environments. They are now a principal target group for the sales of VR headsets, which promise them the immersive experience (chapter 5) they have always sought. *Half-Life: Alyx* (discussed in chapter 1) is a highly detailed postapocalyptic first-person shooter, the first VR title in a well-known franchise that started in 1998 with *Half-Life*. There are others like it. *EVE: Valkyrie—Warzone* (CCP Games) is a futuristic dogfighting game that also exists in PlayStation 4 and PC versions. You are a pilot of a single-seat fighter-style spacecraft surrounded by friendly and enemy spacecraft. It recalls both earlier video games and the fight scenes in films like the Star Wars franchise. *Rez Infinite* (United Game Artists and Sega 2016) is an update of *Rez*, a 2001 rail shooter game, part of a venerable genre in which the shooting figure follows a predetermined track through the world. In the 2016 update, the player can move in 360 degrees, and in the VR version this freedom of movement results in a more vivid sense of immersion. *Lone Echo* (Ready at Dawn) is another screen-based game with a VR version. It has a more cinematic storyline, in which you play an AI robot in the year 2126, whose task is to help your human counterpart as she confronts challenges on a mining station near the rings of Saturn. In *Subnautica*, your spaceship crashes on a watery alien planet, and your tasks are to explore and to survive. The game is doubly immersive as the action takes place underwater. The kaleidoscopic combination of shooter and action mechanics and sci-fi or fantasy storylines will no doubt produce many (hundreds? thousands?) such VR games in the future.

AR shooters or adventure games are less common. But there is, for example, *The Walking Dead: Our World*, part of a transmedia franchise by Next Games that includes the original comic books, the television series, a set of screen-based video games, and a VR game, *The Walking Dead: Saints and Sinners*. In the AR version for a mobile device, GPS is used to determine your location in the world. Your location and the zombies appear as an enhanced Google Maps map on the phone screen. You can also view the zombies and your fellow human survivors as 3-D animations in the AR mode in the game (figure 6.1).

The game is not only a first-person shooter, but also a mapping game, in which the player can walk around her physical neighborhood to reach zombies and other targets. In this sense, it belongs to the same genre as the less violent *Pokémon Go* (Niantic), in which you merely catch Pokémon (rather than shoot them in the head). *Jurassic World Alive* (Ludia), another example of this growing genre, lies somewhere in between *The Walking Dead: Our World* and *Pokémon Go* on the scale of violent engagement. You shoot the dinosaurs with syringes to extract their DNA. All these games use AR to place digital content in the player's physical world, seeking to achieve a close link between the digital and the physical.

Even in this relatively early stage of adoption, VR and AR games are not limited to well-known genres such as shooters and action-adventure. Some remediate games that are aimed at younger players and are cartoon-like rather than aiming at photorealism. *Pokémon Go* featured cartoon-like characters, and yet it was phenomenally successful with adult players. The

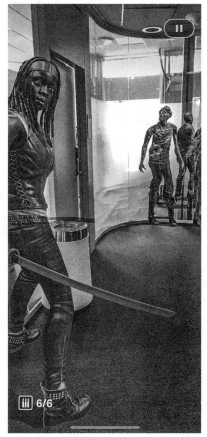

Figure 6.1
The Walking Dead: Our World is a mobile AR game that uses GPS to put the action on Google Maps with an app-specific overlay at your current location (left), but also has an AR overlay (right) that shows zombies and humans in your video feed by using surface detection. © 2018 Next Games.

appearance of a VR platform game such as *Danger Goat* is not surprising. Platform games were originally two-dimensional (*Super Mario Bros.* is probably the most famous); the third dimension was introduced gradually in the 1990s. *Danger Goat* attempts to revive or at least revisit this genre by immersing you in the goat's (game)world.

There are several games that use AR or VR to get the player moving. Traditional (pre-digital) games that require rhythmic physical movements are possibly the source for remediation by the popular VR game *Beat Saber* (Beat Games), a rhythm game in which you use lightsabers (your game controllers) to explode blocks that fly at you to the beat of various songs (figure 6.2). *Supernatural* (Within), designed for the Oculus Quest, is a subscription-based, full-body fitness service for virtual reality, released in the US in April 2020. In both *Supernatural* and *Beat Saber*, although they belong to quite different VR genres, there is an emphasis on the player's body movements.

As for AR, there are numerous casual AR tabletop or floor games (such titles as *Jenga AR*, *Beer Pong AR*, *AR Smash Tanks*) that use surface detection (chapter 4) to place the game play area on a table or on the floor. The play appears to take place in your room (via the video feed of your mobile phone), but otherwise these games simply remediate screen-based titles and genres.

Another genre well suited for the new reality media is sports games. The largest producer, EA Sports, offers a set of franchises, including many major sports, and VR versions seem inevitable. Players of the screen-based versions of these games clearly feel a strong desire to be part of the sport, either as players (e.g., the quarterback calling plays) or as coaches and managers. The games sometimes remediate television broadcasts, with commentary

Figure 6.2
Beat Saber is a VR rhythm game with a little Star Wars thrown in. The player uses lightsabers to keep the beat. © 2018 Beat Games.

on each play provided by sportscasters, and in fact the swooping camera takes the player into the action with more immediacy than the cameras of actual broadcasts. VR can remediate and improve on both television sports coverage and the sports games themselves. As for AR, there are already applications such *NBA AR Basketball*, a simple game in two parts. In the first part, you use surface detection to set up a basket in your living room and then shoot virtual basketballs with a flick of your hand. The ball then flies toward the virtual basket and bounces off the rim or goes through the net with fairly realistic physics and graphics. In the second, you can call up an AR portal and place it on the floor near you. You can walk up to the portal and peer through to see famous basketball moments in 360-degree video, or you can walk through the portal so that the video surrounds you. Both a screen-based sports game and the television broadcast of the sport itself are remediated.

AR and VR can also serve as applications for playful, creative expression, without explicit competition or winning and losing. *Minecraft VR* is a fully immersive, headset version of the sandbox game that already runs on computers, game consoles, and mobile devices. It is called a *sandbox game* because it constitutes a delineated environment in which players can make their own structures and objects out of LEGO-like blocks.

The VR version extends the playful metaverse now inhabited by the creations of over ninety million active monthly players. *Minecraft Earth* was the AR version for smartphones (chapter 1), in which the blocky objects attach to surfaces around you, bringing the digital world in touch with the player's everyday world. *Minecraft* creations persist in a metaverse world divided into many different biomes (desert, jungles, snowfields), and the game can be single or multiplayer. Other drawing platforms or experiences, such as *Tilt Brush* by Google (chapter 1) or *AR Lightspace*, are based on the same sandbox principle. Both allow you to draw persistent lines of light in the space around you. Artist Danny Bittman has used *Tilt Brush* to create virtual landscapes—for instance, in *Aeronaut*, a virtual reality work created for Billy Corgan's music.

To sum up, VR is at the moment the more popular reality medium for video games because it strives to fulfill the dream of immersion that has been a key feature of video games since the arcade days of the 1980s. Both specific games and game genres are easier to remediate in VR. But AR tabletop and geolocated games also figure significantly among current and likely future applications and may ultimately become the more persistently used reality medium of the two, for reasons that we will outline at the end of this chapter.

Entertainment and Art

Video games are of course entertainment, and many say that video games can be art. We treated video games separately simply because they constitute the largest and the most popular digital entertainment form. Even the VR versions of video games are likely to be experienced by far more people than any of the pieces discussed here . . .

Pornography

. . . except perhaps for digital pornography, which is the only genre that can plausibly claim to surpass video games in popularity, although exact figures are hard to come by. In 2018, a conservative estimate for the online pornography industry as a whole was $15 billion (Naughton 2018). To date, VR and 360-degree video porn count for just a small fraction of that whole. Pornography, or at least erotic images, often appear early in the history of many reality media. Erotic photography is almost as old as the daguerreotype itself in the nineteenth century. The first film, or at least one of the first, of a woman undressing, *Le Coucher de la Mariée*, dates from the same year that the train arrived in LaCiotat station, 1896. It is not surprising that 360-degree videos and VR would serve as platforms for pornography as well.

VR porn is already an established genre. The 3-D graphic VR porn takes the form of linear videos, or sometimes participatory video games, but the dominant form is 360-degree video. There are companies that specifically focus on VR and 360-degree productions, such as Reality Lovers and Kiiroo, but the VR category can be found at many of the major pornography distributors—the largest of which at the moment is the Canadian site Pornhub. During the COVID-19 pandemic, Pornhub saw both exponential growth in traffic and increased criticism for unethical practices; see, for example, Cole and Maiberg 2020 and "Pornography Is Booming during the COVID-19 Lockdowns" 2020. The company was finally forced to delete two-thirds of their content—all unverified videos—due to critical reporting (Heater 2020; Kristof 2020).

As with other film genres that appropriate 360-degree filming, the sense of immersion, of "being there," seems to be perfectly suited for the needs of the porn viewer. Such pornographic films are filmed with an actor, usually a male, wearing 360-degree filming gear so that the viewer sees part of his body and shares his point of view. The point seems to be that the viewer should feel as if he is part of the scene (and thus catering primarily to a cis-gendered male heterosexual audience). Of the films that we watched, many were limited to 180 degrees rather than the full 360, and the focus was on female performers' actions and movements. Like the 360-degree documentaries discussed later in this chapter, pornographic VR and 360-degree film remediate the cinematographic and narrative conventions of pornographic films with the added attraction of a virtual environment that supposedly allows for a fuller immersion into the illusion of intimacy. This is at least what the hype and marketing suggest that VR and 360-degree films can offer. The rhetoric is yet another version of the La Ciotat myth—that perceptual immersion enables the user to forget the medium and experience a sense of presence.

AR, VR, and MR Art

The term *art* is now applied to a very wide range of creative practices, but we are here focusing on works produced by the members of the recognized arts community or by those who seek to become part of that community. Digital art in general is an extension of multimedia and installation art dating from the 1960s, the beginning of the postmodern era, when the scope of possible artistic media was expanding to include video and audio technology, along with all sorts of mixed media and found objects. In the late 1960s, for example, several well-known artists collaborated with engineers in a project entitled Experiments in Art and Technology (E.A.T.), promoting the introduction of new media technologies into the art world, such as video projection and wireless sound (Cartiere and Zebracki 2015). In 1970, Gene Youngblood published *Expanded Cinema* (Youngblood 1970), which announced a new field including video art and computer art; the title itself identified these new forms as remediations of cinema. VR pieces came later: one of the earliest examples was Jeffrey Shaw's *Legible City* (1989), which showed a 3-D *city of words* on a giant screen. To experience the work, the user got on a (physical) bicycle and navigated down the virtual streets.

Digital art (and especially digital installation art) has since won increasing acceptance in the larger art community, and increasingly, virtual reality, 360-degree, and augmented reality media art are exhibited at various galleries and museums across the world. Digital artists still often exhibit in their own festivals, shows, or galleries, such as ISEA International, founded in the Netherlands; Copenhagen Contemporary's CC LAB in Denmark; or more permanent collections such as at Ars Electronica in Linz, Austria, and HeK: Haus der elektronischen Künste in Basel, Switzerland. The arrival of specific VR and AR art depended on the successful development of those platforms beginning in the 1980s and 1990s and has flourished recently with smartphones and the newer inexpensive headsets. The work ranges from the popular to the esoteric and political. Much of it belongs to the avant-garde traditions of twentieth-century art. AR and VR art is entering into the mainstream gallery system as well, with gallery sites such as the Kohra Contemporary seeking to be a bridge between virtual reality art and the established art world.

Like video games and 360-degree video, VR art emphasizes immersion as the feature that makes the experience unique—as in a VR work by Christian Lemmerz entitled *La Apparizione* (2017):

> *The gallery is filled with constructed viewing areas, cardboard walls that open up on one of the four sides, forming a safe space for the visitors' VR viewing. The head-mounted VR displays are attached from the ceiling. Once you take one up and put it and the headphones on, you have a certain set circumference as you are tethered to the equipment. You hear a dripping, wet sound. The dripping is too slow for it to be water. It's dark. As you turn around you see—very close to you—a towering golden figure. It is a Jesus figure hanging from a cross. It seems as if he is looming over you, almost leaning over. His body is not static; it seems to be melting, like molten gold dripping down. The sound is louder now, almost syrupy. You see that you can walk around him. He continues to disintegrate in front of you, and although you realize that you are perhaps supposed to think that this is a critical commentary on something (art? religion?), mostly you are physically affected by this animated statue (figure 6.3).*

This work declares itself to be *critical art*, art that critiques art itself and society—in this case, more specifically, religion. There are also works that

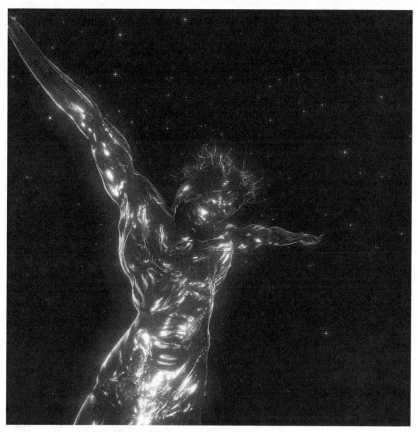

Figure 6.3
A view from within a VR artwork. Christian Lemmerz, *La Apparizione*, 2017. Reprinted with permission by the artist and Khora Contemporary.

exploit VR for the pure joy of immersion. The idea of falling into a computer graphic world was realized in various VR pieces as early as the 1990s, with a CAVE version of Velasquez's *Las Meniñas* (Bizri, Johnson, and Vasilakis 1998), in which the viewer could enter a 3-D reconstruction of the painter's studio as seen in the original painting. In most of these experiences, the emphasis is on realism: the paintings themselves were perspective paintings to begin with. In a series of AR and VR works for the MAK Museum Vienna, the Viennese secessionist artist Gustav Klimt's mosaic frieze in the Stoclet House in Brussels was imaginatively remediated. Frederick Baker and Christian Leiss GmbH created the VR work *Klimt's Magic Garden* (2018). Wearing a headset, the viewer enters into Klimt's aesthetic world. Baker's remediation isolates, multiplies, and animates visual elements of Klimt's original studies and paintings into a garden of graceful forms for the viewer to move through and explore. Cuseum's 3-D remediation of Klimt's *The Tree of Life* (1909) uses Apple ARKit to place the tree in the world (figure 6.4).

Site-Specific Art in Augmented Reality

As we discussed in chapter 5, AR facilitates a different kind of immersive presence than VR. Rather than transporting us to another world, it situates us in a digitally enhanced version of this one, redefining our relationship to our surroundings. AR art pieces therefore strive for different effects than VR art. For example, AR can be used to locate 3-D digital art pieces somewhere in the world and compel the viewers to reflect on the relationship between the art and the location, such as in the Google Arts & Culture app, which has several augmented reality experiences (Google LLC 2016).

Figure 6.4
Klimt's *The Tree of Life*, MAK Museum Vienna by Cuseum. Reprinted with permission.

Given the importance of location and place for AR applications, the link to graffiti and street art is not surprising. By its very nature, street art has a precarious existence. It can be painted over at any time by other street artists or by the owners or the authorities. A digital studio named Heavy used AR to rescue graffiti from this fate in the case of the well-known Bowery Wall in New York City. The Bowery Wall is a palimpsest of graffiti art over several decades. It was first painted in 1982 by the legendary Keith Haring and subsequently by several other street artists. The Heavy project team created an app that allows you to aim your phone or tablet at the current image and reveal any of the earlier layers ("Augmented Reality Mural Resurrections," n.d.).

For the Bowery Wall project, the current owner of the wall itself was a willing collaborator. But AR art does not require the collaboration or even the permission of the owner of the physical space. As long as you can bring your device into the space with the application downloaded or available through Wi-Fi, the art can appear. Not surprisingly, digital artists have exploited AR to invade the space of museums where they could not hope to place their physical work. They have invaded New York's Museum of Modern Art (MoMA) at least twice, for example. The first invasion came in 2010, when Sander Veenhof and Mark Skwarek staged an exhibition called *WeARinMOMA*, in which several artists were able to locate virtual pieces in MoMA's prestigious galleries. The two virtual curators reveled in the transgressive nature of their exhibit. Their website included a photo of a futile sign that MoMA had posted reading: "No augmented reality beyond this point please" (Sterling 2010). Damjan Pita and David Lobser staged another MoMA intervention in 2018, this time a group show called *Open to the Public* that included the artists exonemo, Erin Ko, Manuel Rossner, and Akihiko Taniguchi. The MoMAR app showed the artists' work in the Jackson Pollock gallery. The more stable technology of surface detection allowed them to overlay and remediate the famous drip paintings of the high modernist Pollock (Katz 2018).

These examples all show how AR can establish a critical distance between the art and the physical (or cultural or historical) location. In VR, you fall into the artwork; in AR, you are made aware that the work joins you in your space and perhaps has not even been invited there. Augmented reality artworks often rely on a sense of space and place, but they also can function as the medium for a 2-D or 3-D work that could be shown in other ways. Artist

Tamiko Thiel's augmented reality works exemplify different strategies for how to use AR, such as the place-specific *Treasures of Seh Rem* (Thiel 2017), which was part of the Boston Cyberarts commissioned work for the *Augmented Landscape* exhibition in Salem, Massachusetts. Thiel's work *Strange Growth* (Thiel with /p 2019) directly links the digital portion of the work to the printed pages of the *VECTOR* magazine, and using the position of the user to show curiously shaped, multidimensional gray figures. In addition, there are AR works in which the relationship to the place where they are shown depends on the user's impression of them. Such works often use surface detection to place virtual content in the user's immediate environment, at times with surprising effect. During 2020, *Acute Art* (directed by Daniel Birnbaum) released an AR app that displays 3-D work by artists such as Olafur Eliasson, KAWS, Christo and Jeanne-Claude, and Cao Fei wherever the user chooses (Acute Art 2020). The artworks include 3-D animations of a burning sun or a rain cloud with rain drops that hit your floor (Eliasson), and a young boy sitting by a table drinking a soda—who, when you get closer to him with your device, asks if you have seen his dad (Cao Fei). There emerges a connection between you, the place, and the artwork, in part because the works invite you to walk around and look at them from all angles; you can even photograph or videotape them. The connection will be different for each person who uses the app.

Mixed Media Experiences

VR and AR art emerged at a time when the art community had already rejected the idea of *medium purity* that characterized much of earlier modernism. Dick Higgins and the Fluxus artists championed *intermedia* combinations of objects, poetry, performance, and music, and many other postmodern artists and movements followed suit, fashioning their own combinations. Art could be made out of anything material. In the 1980s and 1990s, digital technologies came to be added to the mix and became part of installations. Today, mixed experiences combine VR, 360-degree video, or AR with physical components and live actors, as does an experience by the Danish team Makropol called *DoomRoom*:

> *You are part of a viewing party who have all bought tickets to be part of the mixed media experience* DoomRoom. *The show starts when a young*

woman, completely naked and covered in white paint, brings you up to an anteroom. There, she performs a theatrical start, serving as a gateway into the main part of the show, which consists of watching a series of 360-degree films with a headset as you are led by hands that touch you and gently nudge you here or there through a physical space that you cannot see. Some of the films have an eerie feel to them, with some clear Lynchian aesthetics or surrealist elements. In one film, you find yourself in a pink bar with a bartender who is so tall that his head grazes the ceiling. Suddenly, you sense that someone is standing next to you. As you turn to look, you can see that two men with naked, muscular torsos have appeared on either side of you, standing very close to you. The final film ends with the images of a young man walking toward you. He comes closer and closer. When he is standing in front of you, he stretches out his hand. You stretch out yours, and as you do a human hand takes hold of yours, ever so lightly. Standing there holding hands, someone takes off the head mounted display and headphones. Now you see that in front of you in this real space—a darkened room with black scrims hanging all around—stands the young man you just saw in the film. He smiles, whispers "thank you," and leads you out of the room.

Digital artists do not necessarily share the popular desire for total VR. They are often as concerned with the physical space of their works as with the virtual space. How should they configure that space for the participant? Should she sit down and only see a 180-degree experience? Should she be able to walk around? How can the space be delimited so that she will feel safe moving around? Or should the experience challenge any sense of safety? The Swedish performance art company Bombina Bombast suggests that this contract between the artist and the audience should not preclude experiences that are challenging or unexpected for the audience members (Emma Wexell and Stefan Stanisic, pers. comm., March 13, 2018). In their work, such as *Fritt lopp för de dugliga/Horror over Dalecarlia* (2015), Bombina Bombast has worked with incorporating both expected and unexpected physical elements while the audience member is viewing the work in a headset. In Makropol's *DoomRoom*, the physical encounter with the young woman at the beginning is, in jargon borrowed from management theory, an *onboarding* experience, and the meeting with the young man at the end is not an afterthought, but rather a physical culmination of a 360-degree cinematic experience set in an unknown physical space.

Fictional Storytelling in Reality Media

Film and television drama remain our culture's most influential forms of visual storytelling. Many enthusiasts hope and expect that VR will develop into a compelling narrative medium as well, eventually surpassing film. Yet the very idea of VR storytelling is perhaps the clearest example of remediation, where the models for VR are film, television, and live theater. VR and 360-degree video (again, the two are not always distinguished) are already occupying special tracks at film festivals, including Tribeca, Sundance, Cannes, and the International Documentary Film Festival Amsterdam, to name a few. During spring 2020, when COVID-19 forced many film festivals to place their content online, immersive experiences were also showcased online. Tribeca Virtual Arcade, Cannes XR, and the Museum of Other Realities collaborated to offer an online exhibition during a few days in June and July 2020 (Cannes XR, n.d.).

What the proponents of VR and 360-degree video storytelling imagine is a form that adds to the emotional impact of film, particularly through an enhanced feeling of sensory immersion. This claim of immersion has been made too for the linear (noninteractive) experience of 360-degree video, which is already a recognized medium for short narrative films and documentaries. Like traditional film, especially at the festivals, 360-degree videos already run the gamut from the recognized popular genres to the experimental. A professionally acted and scripted example of a popular genre piece is the Dutch, gangster-themed, short 360-degree film *The Invisible Man* (Keijzer 2016). You as the viewer are the titular invisible man, who sits at a table with three seemingly criminal types as they play Russian roulette. At the end of the drama, and with an ironic emphasis on immersion, one of the gangsters picks up the gun, points it at you, and shoots.

Examples of films that forgo conventional narrative structures or emphasize other forms of storytelling include *Through You*, an experimental dance project by Lily Baldwin and Saschka Unseld (2017) that portrays two lovers over a lifetime, expressed by dancers' movements. The virtual reality works *Queerskins* (2018–2020) by Illya Szilak and Cyril Tsiboulski are also examples of how a new medium can engage viewers in unexpected or experimental ways. The *Queerskins* stories explore the complex relationship between a devoutly Catholic mother and her gay son who has died of AIDS.

There are growing repositories of 360-degree videos, such as Within, Blend Media's 360-degree stock photos, and of course YouTube. The vast majority of current videos on Blend Media and YouTube are personal amateur productions, often shot with a GoPro or similar inexpensive wide-angle or panoramic camera. Such videos cover a variety of subjects: personal adventures (a scuba dive or skydive, a ride in a helicopter, or a hike through a national park), a sporting event, a moment in nature (an impressive sunset). Most have only the simplest story to tell, if there is any story at all. We could class them in their own genre as actualities, like the *actualités* (including *La Ciotat*) that were the first films by the Lumière brothers at the end of the nineteenth century. But Within and other sites contain professionally produced 360-degree videos in a variety of familiar genres, including horror and animation. Horror is a natural choice for this medium because it can exploit the fact that you as the viewer need to be looking around constantly: the monsters may be coming from any direction. Horror also emphasizes the uncanny, which suits the medium, as we have argued. Another obvious choice is animation, with its emphasis on stories told in a visually simpler mode rather than through the elaborate dialog and acts of dramatic live action. Beyond these genres, some well-known filmmakers are beginning to experiment with 360-degree video or true VR. Alejandro González Iñarritu, the director of *The Revenant*, made *Carne y Arena*, the first VR piece to premier at the Cannes festival in 2017. And Brett Leonard, who in 1992 made one of the first major Hollywood movies *about* virtual reality, *The Lawnmower Man*, made a 360-degree movie, *Hollywood Rooftop*, in 2018. In Leonard's words, it is a "transitional hybrid cinematic immersive piece" (Bye 2018).

Most of these examples of 360-degree video storytelling are linear, like traditional film, with or without elaborate cuts. In most cases, the only difference is that the viewer can choose to look in all directions while the action plays out. It is possible to make a 3-D VR experience run in the same linear fashion. The action would play out as with 360-degree video, but the viewer would have six degrees of freedom rather than three (chapter 4). This would be prerecorded VR, like 360-degree video. True VR, in which the 3-D models are being redrawn and manipulated in real time, offers the potential for interactive narrative—a kind of storytelling in which the viewer somehow becomes a participant in the action—but this also requires higher performance from the VR systems. Interactive narrative in a

VR environment has been the dream of enthusiasts since at least the 1990s (Murray 1997). It would be like the holodeck of *Star Trek: The Next Generation*. But a digital narrative that combines a sophisticated dramatic storyline (remediated from film) and effective participation of the user has rarely yet been achieved, with or without the 3-D graphics of VR. One sophisticated example was the AI-based interactive drama Façade (Mateas and Stern 2003), which used a multi-level story manager to allow the participant to interact with two computer characters using natural language, bending the story to one of a half-dozen outcomes based on their actions. Even without embracing fully AI-guided narratives, some genres of VR video games today (e.g., first-person shooters such as *Half-Life: Alyx* or *Pavlov*) do contain narrative elements. What about AR? It suggests the possibility for storytelling that is based on the user's location. As the user moves through a place, she could hear appropriate narrative and even see characters. The Blast Theory group of interactive artists has been experimenting with location-based storytelling for over two decades (Blast Theory 2019). The stories have been fashioned either by Blast Theory or by the participants themselves as they leave audio traces for other participants. Façade was converted into an AR-based experience, putting the participant in a see-through AR display inside a space made to resemble the original virtual apartment in the screen-based version. A behind-the-scenes "Wizard of Oz" interpreted their actions while still allowing the story engine to control the experience, giving a participant the freedom to move anywhere, interact with objects in the space, and say what they wanted (Dow et al 2007). However, most of the current storytelling applications in AR and VR have not been in the realm of art or fiction, but rather in journalism, documentary filmmaking, or cultural heritage applications.

Documentary and Journalism

In November 2015, the *New York Times* released *The Displaced* (Solomon and Ismail 2015), a collaboration between Within (then VRSE) and the *Times*'s new initiative, the VR platform nytvr (now defunct). Along with this release, to all its subscribers, the *New York Times* also shipped Google Cardboard, a simple headset that turns a smartphone into a VR viewer. Branded as VR, *The Displaced* was in fact a 360-degree video documentary. The film borrowed from existing strategies in documentary filmmaking, including on-location filming in South Sudan, Ukraine, and Syria. The producer, Chris Milk, had already made a 360-degree documentary together with Gabo Arora with a similar storyline—children suffering in war-torn regions—called *Clouds over Sidra* (Arora and Milk 2015), which we discussed in chapter 5.

In both *The Displaced* and *Clouds over Sidra*, the camera is an invisible center, the viewer's point of view. It is an unframed experience, unlike a documentary shot with a camera that frames the scene, although the direction of the viewer's gaze is often implied by creating sites of attention with the filmed subjects or sounds. In many 360-degree films, including these two, the camera has been edited out of the shot. Looking down, the viewer sees neither the camera nor the cameraperson's body. Making the camera disappear is not always easy, as becomes apparent in *The Displaced* when a child in the film takes hold of the stand on which the camera is mounted and we see the camera's shadow on the ground as the child moves. People moving closer to the device become close-up shots, whether intended or not. Camera angles, mise-en-scène, and the staging of the scenes contribute to our viewing experience just as they do in conventional framed films. These 360-degree videos are recognizable as remediated documentaries.

This genre is growing. An online database of both 360-degree videos and 3-D VR, discussed in this chapter, contains over six hundred examples to the end of 2018 (Bevan and Green 2018). The *New York Times* collaborated on further examples, such as *Man on Spire* (Chin and Solomon 2016), about a man who climbed the building at 1 World Trade Center. CNN has an archive of news vignettes shot as 360-degree videos; although again branded as VR, these are mostly live action with a few animations. The *Guardian* uses 360-degree film in its VR app for its immersive journalistic pieces, such as *The Party* (Bregman, Fernando, and Hawking 2017), which allows you to hear the thoughts of a young woman on the autism spectrum as she attends a birthday party. *National Geographic* has created an entire YouTube channel of 360-degree videos on traditional nature subjects, such as lions, sharks, and orangutans, that follows the same uncontroversial, narrative pattern as its television series. In its other media forms, photography and television, *National Geographic* has always focused on spectacle, so it is not surprising that it has turned to the immersive spectacle of 360-degree film.

In addition to 360-degree documentaries, there are experiments in documentary film in 3-D VR. The documentarian Nonny de la Peña has worked for years in this area. Along with collaborator Peggy Weil, she remediated her own film documentary, *Unconstitutional: The War on Our Civil Liberties* (2004), into a first-person experience, *Gone Gitmo* (2007). Realized in the virtual world *Second Life*, *Gone Gitmo* places you in a prison cell and exposes you to torture techniques. The later *Hunger in Los Angeles* (2010) uses a VR headset to immerse you in the experience of waiting in a food bank line (figure 6.5). Her production group Emblematic has since created a number

Figure 6.5
Nonny de la Peña's *Hunger in Los Angeles* (2010), an early immersive documentary. Image by Peggy Weil. Reprinted with permission.

of VR experiences, some of which have appeared at Sundance, attesting again to the remediating relationship between film and VR. De la Peña calls this work "immersive journalism" and, like Chris Milk, has argued that her VR pieces foster a particularly powerful empathetic reaction because the user feels present (de la Peña 2015).

De la Peña's VR work falls somewhere between strict fidelity and fictionalized documentary. Perhaps VR as a medium invites this kind of creative license because 3-D graphics does not convey the same sense of realism as photography. Other VR pieces also fall in this space between. Gabo Arora and Saschka Unseld's *The Day the World Changed* (2017) combines the testimony of survivors with photogrammetry and computer graphics to create a virtual experience of the effects of the atomic bombs dropped on Japan in 1945. Although it was shown at Sheffield Doc/Fest, a documentary festival, in 2018, the work is an artistic mixed media work as much as it is a documentary.

AR documentary is less well developed than VR, both because it is harder to argue that AR is an empathy machine and because AR does not obviously remediate film with its long documentary tradition. But online versions of major newspapers such as the *New York Times* and the *Washington Post* are defining a genre of news article that employs AR (along with VR) as a new mode of illustration, representing a remediation of journalism rather than film. As part of its digital strategy, the *New York Times* has offered a series of feature articles to use augmented reality "in a journalistic way" (Snyder 2017). Features on David Bowie's costumes (chapter 5), a visit to a cave in Thailand, the *Insight* Mars mission, and the athletic moves of four 2018 Winter Olympians (New York Times 2018a, 2018b; Snyder 2017) all follow the same format. Each consists of a web page with elaborate graphics that you can read on your computer or on a mobile device. An image on the web page serves as a portal into the AR experience. If you are reading on a tablet or phone, you can use surface detection to set up stable virtual objects (the costumes, the Olympians, the *Insight* lander, and so on) on your floor and can walk around and explore them. These articles promote AR as a form of respectful remediation of the traditional feature article. VR experiences could be embedded in articles in the same way. As the standard browsers (Chrome, Firefox, etc.) improve their support for the immersive web, we can expect more of both formats, at least as journalistic special features. During the spring of 2020, during lockdowns and self-quarantine, various VR and 360-degree media experiences received even more attention from people who missed being able to visit museums and concerts or walk around freely or travel (Feinstein 2020).

There are also AR stand-alone documentaries. Asad J. Malik's work *Terminal 3* (2018) is an interactive, augmented-reality documentary that explores how Muslims may experience US airport interrogations. However, you are not the one being interrogated; rather, you are the one who is asking the questions and ultimately decides who will be let in and not. The work is designed for HoloLens 2, and the persons in front of you appear as holograms that represent actual people and their experiences (Mashable 2018). *Only Expansion* by Duncan Speakman (2019) is an audio experience that uses "real-time modulation of the surrounding soundscape mixed with produced sounds" (Bye 2019), creating a site-responsive augmented audio experience.

Museums, Cultural Heritage, and Tourism

Both VR and AR have considerable potential for staging experiences in museums, cultural heritage, and tourist sites, and some of that potential has already been realized (Bekele et al. 2018; Jung et al. 2016). Museums and heritage sites are committed to bridging the gaps between now and then or here and there—bringing the past into the present or bringing what is physically distant into the visitor's presence. If you are standing on the Acropolis in Athens, you see in front of you the ruins of the Parthenon, the temple to the patron goddess Athena, and other temples and structures, such as the Erechtheum and the Propylaea. You have to imagine what the whole hill looked like in 400 BC. If you are standing in the British Museum in London, looking at the Elgin marbles, you see fragmentary sculptures that were once part of the Parthenon, 1,500 miles away. That distance has historical, cultural, and political implications.

AR and VR cannot do anything to solve the political dispute between the UK and Greece, but they can reunite sites and artifacts virtually, in both time and place. If you are in the British Museum, you could call up on your phone a VR reconstruction of the exterior of the temple, which would allow you to see the sculptures as a frieze extending around the top (figure 6.6). If you are standing on the Acropolis, the same 3-D VR reconstruction appearing on your phone would allow you to compare the present ruins

Figure 6.6
Two views of a 3-D model of the Parthenon: (a) outside and (b) inside the temple.
Images by Colin Freeman.

with the intact original. It would permit you to move into the interior of the temple and view the statue of Athena Parthenos. A visitor today is not allowed inside the ruins, and in any case the statue was destroyed or taken away at some time in the first millennium AD. The notion of presenting

VR reconstructions to visitors of historical sites has been named *situated simulation* by Gunnar Liestøl, who has prototyped such experiences for the Parthenon, the Roman forum, the Via Appia, and the Viking Ship Museum with its ships interred in burial mounds outside Oslo, Norway (Liestøl, n.d.). Others have remediated historic exhibition spaces as showrooms for virtual exhibitions, as is the case with *Thresholds*, a VR recreation of Henry Fox Talbot's *Model Room*, one of the world's first photography exhibitions in August 1839 (Tennent et al. 2020). The VR experience involves exploring a virtual and a physical space, which allows the visitors to touch the physical environment while seeing a digital representation that is textured onto the physical surfaces. This creates a situated experience, an overlay or simultaneously here-and-there feeling. Therefore, the physicality of being in an actual museum space is acknowledged and referenced in the digital VR room.

There have been many experiments with using AR overlays to bring the past into the present location (Bekele et al. 2018; Billinghurst et al. 2015). In the 1990s and 2000s, AR researchers worked to develop the techniques required to overlay images and objects exactly in a location, and they succeeded only under special conditions and with expensive technology. Although the problem is still a hard one, surface detection and image and scene recognition have now come to the consumer level of mobile devices (chapter 4), which is particularly valuable for AR applications for museums and heritage sites. In chapter 5, we discussed a prototype that allows the visitor to see a reconstruction of a building called the Middle Stoa in the Athenian agora (the marketplace located below the hill of the Acropolis where the Parthenon stood). With this geolocated AR application, the visitor could hold up the phone and see the present ruins with the Middle Stoa overlaid. These kinds of virtual reconstructions are becoming easier and more common with each new generation of mobile devices.

Paintings or objects in museums can be recognized and enhanced in various ways, although professional curators remain skeptical of anything virtual that might get between the visitor and their museum pieces. The applications of AR in remediating museum exhibits continues to grow; examples include an augmented tour at the Morgan Library in New York City; a Viking tour in the Royal Alberta Museum in Alberta, Canada; an art piece in the hanging gardens of the Pérez Art Museum Miami; and an AR experience to commemorate the moon landing at the National Air and

Space Museum in Washington (Levere 2019). A Boston-based company, Cuseum, provided a particularly ingenious example of bringing back the past in the Isabella Stewart Gardner Museum in Boston, from which thirteen highly valuable paintings were stolen in 1990, including works by Rembrandt and Vermeer. The paintings have never been recovered, and the museum continues to display empty frames to commemorate the loss. In 2018, programmers at Cuseum had the idea to restore the paintings virtually. Surface detection made it possible to set two stolen Rembrandts stably in their frames without the jitter that plagued earlier AR (figure 6.7). The museum was not involved in this virtual restoration, by the way, and it was not made publicly available (Cascone 2018; Katz 2018).

All these VR and AR applications for cultural heritage and museums can be understood as remediations of earlier forms: human-guided tours, audio tours, printed museum guides, photographs, films, and even postcards made for tourists and visitors. They offer new ways to convey the aura (chapter 5) of places, artifacts, and works of art that our culture as a whole or some particular cultural community regards as special.

Figure 6.7
The Hacking the Heist application by Cuseum restores stolen Rembrandts virtually to their frames at the Isabella Stewart Gardner Museum. Reprinted with permission.

Visualizing, Learning, Training

Since the earliest days of AR research in the 1990s, researchers have imagined that the most important applications would be not be games or social media (they did not yet exist) but task-based applications—to help mechanics repair machinery, to help architects and planners visualize buildings and cityscapes, to aid doctors in treating patients, and so on. Such applications are now developing, though at a slower rate than originally envisioned. Even though such task-based applications are not focused on entertainment and in some cases not on communication, they are still forms of mediation. They give workers new ways of rendering their tasks more clearly and easily through graphics, videos, images, sounds, or text. They put information in some media form in the appropriate location. They make use of VR and AR as reality media to redefine the physical and material world that they interact with. And in most cases, they remediate some earlier medium that was used for that purpose.

Most task-based VR and AR involves some form of visualization.

Healthcare and Medical Applications

Healthcare is likely to be of growing importance for AR and especially VR in coming years (Marr 2020). The medical applications generally fall into three broad categories: medical imagery, patient therapy, or medical training. There have been numerous preclinical trials in using AR as an aid for orthopedic surgery (Jud et al. 2020). A HoloLens-based system has been approved by the FDA in which CRT scan information is actually overlaid on the patient's body to aid the surgeon in planning an operation (Novarad, n.d.), which is in fact the same genre of application as HoloGARAGE, the experimental HoloLens 2 application for car repair (chapter 5). In this case, the doctor wears the AR device, but patients are also experiencing AR and VR directly for certain kinds of therapies.

Since the 1990s, various VR systems have been developed to treat patients with phobias, such as fear of heights, or with PTSD, through a technique known as *graduated exposure therapy*—or in this case, *VR exposure therapy* (Boeldt et al. 2019). Bravemind, for example, exposes PTSD patients to traumatic combat situations in a controlled step-by-step manner (USC Institute for Creative Technologies 2005–present). The therapist uses the system to mediate the patient's confrontation of a painful past reality. This approach depends on the capacity of VR as a medium to achieve sufficient presence to evoke the patient's fear, as in the origin story of the La Ciotat train. Simultaneously, of course, the patient and the therapist know that they can break the illusion again at any time. Another aspect of VR presence, which Lombard and Ditton (1997) identified as *presence as transportation* (chapter 5), is being experimentally used as a treatment for pain. Patients experience less pain when they are offered a variety of virtual experiences, including "guided relaxation, natural environments, simulated flights, and animated games" (Young 2019). VR is also being extensively studied in rehabilitation therapy—for example, to help patients improve their mobility as they recover from spinal cord injuries (Yeo et al. 2019). Although VR may ultimately prove to be more effective in all these applications, earlier media have similar therapeutic uses. Videos are often used to show patients how to perform various exercises for physical therapy. Music, film, and television can distract audiences from physical pain. Here too, VR is part of a longer media tradition.

Architecture and Design

Architects have been employing 3-D CAD systems for decades to visualize their designs for buildings and built environments. It is an obvious step to put those models into an immersive VR environment, both for their own work and to communicate their designs to clients—again, because of the development of inexpensive VR headsets. In the auto industry, Ford's Immersive Vehicle Environment (FiVE) has experimented with VR for several years, permitting designers to walk around and sit in CAD models of new designs (McIntosh 2017). BMW and VW have similar uses of VR in design (BMW Group 2018; Volkswagen, n.d.). In architecture and construction, in both VR and AR applications, the display and manipulation of 3-D objects have been most important as they allow for reality media to be used as part of the workflow of the architect or designer. Building information modeling (BIM) systems and CAD models can be displayed via AR superimposed over the real world. Such spatial visualization remains the key gain that both AR and VR systems can offer. Today there exist a host of possibilities to display AR—for instance, by way of specific applications geared toward architecture, such as Fuzor Mobile and ARki. The opportunity to scrutinize 3-D models by way of a VR or AR application is also a tool for displaying designs to potential customers. AR has the added benefit of correctly placing such 3-D models of buildings on the real sites at which they will be constructed. AR applications can also provide an animation of various stages of construction over time so that different systems of a building can be displayed separately—for instance, with documentation from the construction phase that can be accessed once the building is finished. Reality media used in this way allow architects and designers, and their customers, potentially richer opportunities to evaluate the physical, functional, and experiential dimensions of a construction.

Product Display and Retail

Beyond architecture, product and service design is a growing application area for both AR and VR, as is apparent from the auto industry, where companies are using reality media both as part of the design process and to display designs to customers (Seppala 2018). For years, real estate sites have offered panoramic tours of houses in 360 degrees, but usually as static images rather than videos; viewing such tours in headsets is now becoming possible. IKEA's AR app *IKEA Place*, which allows you to visualize furniture in your living room before buying it, is one of a growing number of remediations of the catalog or brochure that retailers have offered since the advent of consumerism in the late nineteenth century. Advertising has emerged as a popular application for AR, particularly the use of image recognition in magazines or printed brochures. The consumer points her phone at an image in a magazine—say, of a new automobile—and the screen displays a 3-D model of the automobile sitting on the image.

Training and Education

In the first wave of euphoria in the 1990s, enthusiasts claimed that VR would become a general learning environment, replacing textbooks in schools. That has not happened (just as it never happened for film and television in the twentieth century), but specialized training and simulation are promising application areas. Physical flight simulators, which replicate the cockpit of the plane, long predate VR and are still important in the military and commercial airline industry. They have gradually assimilated the characteristics of 3-D immersive graphics while maintaining the physical controls of the cockpit. Throughout the 1980s, in the early days of VR, the US Air Force funded the Super Cockpit research program to develop VR hardware and software to train fighter pilots (Bye 2015). The army fielded the first fully immersive simulator for combat soldiers in 2012 (Mer 2012). And since then, the military has continued to develop "synthetic training environments," which combine all sorts of digital technologies, including VR and AR (Fusion, n.d.; US Army, n.d.). This is an obvious step as advanced military combat is itself becoming increasingly digital. While the military has devoted some of its massive resources for decades to funding research into VR and AR for combat and training, the advent of inexpensive headsets has meant that companies such as Walmart, Verizon, and United Rentals can also begin to experiment (STRIVR, n.d.).

Navigation

Google Maps, Street View, and Earth are already on their way to constituting a digital mirror of our world, as we'll discuss in chapter 7. Google Earth can generate an immersive VR bird's-eye view, and Google's VPS aims to overcome the key problem that, as we noted in chapter 4, has plagued outdoor uses of AR for years. Using scene detection and crowdsourced visual material, this system is meant to achieve "accurate orientation: similar to humans, it looks for visual cues like the facades of landmarks and stores to figure out where you are" (Google Developers 2019). The system, now active in Google Maps as the Live View feature, will allow users to see directional arrows in AR to help them determine the directions indicated on the flat maps on their screen (figure 6.8). Already, the photographic mapping of the world in Street View provides a rich source of visual cues that Google Maps can mine for information. This mapping and systematic data analysis of the world also raises issues of privacy that we will address further in chapter 9.

In addition to this general mapping solution, there are and will continue to be apps that help visitors navigate in specific locations and for specific purposes. Gatwick Airport's app, for example, offers an AR navigation feature, which, unlike GPS, functions indoors with the aid of two thousand beacons stationed throughout the terminals (Gatwick Airport Limited 2017).

Social Spaces

VR is often promoted as the ultimate reality media, surpassing earlier media in its capacity to fashion a world for us. It is ironic, then, that the

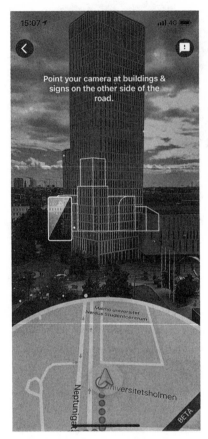

Figure 6.8
Google's Visual Positioning System (VPS) (left) turns Google Maps into an AR navigation system (right). © 2020 Google.

early uses of VR environments include viewing traditional films or playing screen-based video games. Launching the Microsoft Mixed Reality system, for example, conducts you to a villa on a mountaintop. There you find a giant screen, which is your portal to surfing the web or watching videos of movies and television series. It surprised (and dismayed) many when Facebook bought Oculus in 2014, even before the start-up company had released a production version of its VR headset. Facebook saw VR as an opportunity to remediate social media and subsequently developed Spaces to add to its universe of platforms (from watches to smartphones to tablets to computers) for social interactions. The unsuccessful Spaces has been superseded by Facebook Horizon (Stein 2019), but Facebook has not given up on its vision of a social media metaverse (see chapter 8). The VR game emporium Rec Room went online in 2016 (Rec Room Wiki, n.d.): you enter into a virtual lobby, where various games are located behind doors. The lobby and its games support multiple players as cartoon-like avatars, and you can communicate with others as you play.

Mozilla's Hubs features rooms and simple 3-D avatars, but it dispenses with the games. The emphasis instead is on verbal communication, and the visuals are largely to provide a sense of social presence. As of 2020, other major services include VRChat and AltspaceVR. As an example of how services in the field may shift, *High Fidelity*, the VR successor to the screen-based *Second Life*, seemed to take the spatial metaphor further with more photorealistic avatars. In 2019, however, *High Fidelity* switched to an "enterprise model" rather than becoming a mass user platform (Hayden 2019). During the COVID-19-filled spring of 2020, when many found themselves working from home, *High Fidelity* released a web-based 3-D audio service in a 2-D browser environment for social gatherings. The services come and go, but the genre persists.

While VR applications for communication are further along in development, there are AR versions as well. Since 2016, Microsoft has been demoing and promoting an application for its HoloLens device called Holoportation. Each participant is in her own physical space, and her movements are being captured by several cameras. The motion capture data is radically compressed and sent to the other participant's HoloLens as a photorealistic avatar that appears to share her space. Another (obviously much smaller) company, Spatial, is developing a multiplatform version of augmented collaboration, in which participants can share the same virtual space whether they are wearing a HoloLens or working from a variety of other platforms, including an Oculus Quest, mobile device, or computer.

It seems likely that the VR video game environments like Rec Room will be successful because the screen-based versions of multiplayer games like *World of Warcraft*, *Fortnite*, and *Overwatch* are already tremendously popular. The success of the VR environments for the viewing of films and videos will depend in part on the future development of higher-resolution headsets. At best, these environments will be an added feature for those who have bought headsets for other purposes, such as to play VR video games.

The real attraction is that users can watch films together with friends who have their own headsets and may be far away. The idea of VR as a social space, a place to hang out and "connect" with friends, is also what drew Facebook into this medium. It remains to be seen whether the appeal of VR social media will be great enough to compel people to buy and use headsets or whether it will always be a secondary activity of VR gamers, and whether the new VR and AR communications environments can rival or displace real-time videoconferencing, either for business or social purposes. For social communication (in other words, Skype, FaceTime, or the other video-streaming apps on computers and mobile devices), VR seems like a supplement, rather than a replacement. For business communication, VR conferencing with avatars offers a chance to enhance collaboration—for example, for product design. The lockdowns during the COVID-19 pandemic spiked interest in large-scale virtual conferencing, which had been growing even before the spring of 2020 among sophisticated technical communities such as computer scientists and engineers. There was a successful experiment at the ACM UIST conference (on computer interface design) in October 2019. And in March 2020, the IEEE VR conference (on virtual reality research, appropriately enough) was hosted entirely online, including in virtual rooms of Mozilla's Hubs (figure 6.9) under the co-direction of one of this book's authors (MacIntyre 2020). This experiment showed that a hybrid approach involving multiple channels (Twitch, Slack, and Slido, as well as Hubs) could satisfy many of the participants' needs for intellectual and social exchange.

The Filmic Eye and the Body

Many of the genres discussed so far (certainly porn, 3-D storytelling, documentary, and some kinds of video games and art) are defined, to a greater or lesser degree, by their remediation of film, and as in traditional film, the relationship between the camera and the viewer's eye (and body) is a key to the experience; see, for example, Bevan et al. 2019. Director Brett Leonard has said that one of the key differences between ordinary and 360-degree filmmaking is that we are visitors to these worlds, part of a *storyworlding* (Bye 2018). Unlike the avatars that stand in for you in VR or computer graphics worlds (think of massively multiplayer online role-playing games like *World of Warcraft* or the virtual world *Second Life*), these environments

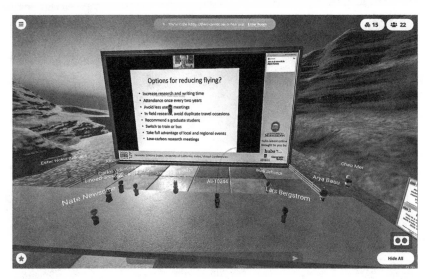

Figure 6.9
IEEE VR's main conference room in Hubs.

promise the experience of being inside a cinematic world. As noted earlier in this chapter, videos in 360 degrees often seek to delete all traces of the camera gear from the film. When you look down in *Man on Spire* (Chin and Solomon 2016), which shows filmmaker and mountain climber Jimmy Chin's ascent to the top of 1 World Trade Center, you do not see a body to whom the viewpoint belongs. The camera rig is effaced from the film so that you see only the dizzying view of Manhattan below (figure 6.10). On the *New York Times Magazine* webpage about the film, an editor's note reads: "Images of the V.R. apparatus and its shadow have been removed in some places. This editing, which is common in V.R. production, helps preserve the scene as a viewer on location would normally see it" (New York Times Magazine 2016). You are invited to see the scene as if you were there, unmediated by film.

The "floating eye" that takes in the scene is a common choice in 360-degree video. Unlike VR pornography, in which the presence of a body is central to the experience, professionally produced 360-degree videos often choose to omit a virtual body. They have this in common with the mapping application Google Earth VR, in which the third-person POV and the modes of interaction suggest a god-like position from which you can fly

Figure 6.10
A still from the 360-degree film *Man on Spire* by Jimmy Chin and Ben C. Solomon.
Photo by Jimmy Chin. Reprinted with permission.

over the earth, zoom down into landscapes and cities, and stand on street corners with the help of Google Street View mode. You can even manipulate the earth itself, spinning it around to visit various parts. The fact that you do not then have a fixed position or the visual traces of a body probably contributes to nausea for some VR viewers.

By contrast, including a headless body in the shot is an aesthetic choice that allows filmmakers to create a link between the actor whose body you see in the virtual world and your own embodied self. This can lead to a jarring or uncomfortable sensation as the body you see is obviously not your own. This representational choice opens up possibilities for various interpretations or reactions. You can be invited to engage with or accept the virtual body as your own, or you can feel a disconnection because it does not correspond to what you are willing or able to embody. In this regard, a male body for a female viewer of VR pornography can present such a defamiliarizing effect.

The Special and the Everyday

Another way to classify all these genres is in terms of the special and the everyday. As we noted in the introduction, our two reality media are different in this respect. The experiences of 3-D VR and 360-degree video are by their natures special ones. They are like film in the sense that you make a conscious decision to devote a couple of hours to watching a film and perhaps go somewhere special—that is, to a theater. VR is almost always experienced that way. You decide to play a VR video game, sit down or go to a space that is clear of obstacles, and put on your headset. With the current generation of headsets, and for years to come, you can't participate in fully immersive VR while walking down the street. (At some point in the future, you might be able to rely on a VR headset's external cameras and sensors to steer you safely amid people and through traffic; see chapter 10.) AR, however, can be for everyday purposes (following directions to a destination) or for special purposes (cultural heritage AR, games). This is more like television, which can be special or everyday: sometimes you devote an hour to a program, but you can also just have it on in the background and monitor it.

The everyday uses of AR can themselves be divided into two categories, which we might call *intentional* and *peripheral*. When you rely on an AR application—say, Live View in Google Maps—to guide you to a museum in a city you are visiting, your use is goal directed. You may not give the app your full attention—you can look around at the street until you have to follow the next direction it provides—but you chose to launch the app for that purpose, and it returns frequently to your mind until you have reached the museum. But an always-on AR app could function at the periphery of your attention. Your phone or someday your AR glasses could be set to remind you whenever you pass by a certain kind of restaurant because you want to check out restaurants where you might dine later. This would be an AR extension of the alerts that you can already set on your smartphone.

At present, the experiences of VR or 360° video are special ones; AR experiences can be everyday or special.

We might also ask in which media forms and genres the La Ciotat effect is strongest. It seems obvious that we will be most astonished the first time we experience a new medium. The point of the La Ciotat legend is that the audience has never seen a film before. The first time or first few times that a user puts on a VR headset, she may be astonished at the immersive visual experience. AR, especially AR in a smartphone, is perhaps less astonishing, even for a first-time user, because we are already accustomed to graphics on the phone screen. But a new app or a new genre can elicit a feeling of pleasant surprise. These feelings will necessarily diminish as we become more familiar with the technologies and the genres. The makers of smartphones and new applications of all kinds try to revive our astonishment. According to Apple, every new model of iPhone has amazing new features; each new installment of a first-person shooter offers amazingly lifelike graphics or detailed storyworlds. Obviously too, special uses of the technology are more likely to be surprising than everyday uses. Immersive VR experiences and AR art are more likely to provoke surprise and a feeling of the uncanny than an AR navigation guide.

In 1946, the film historian André Bazin wrote an essay entitled "The Myth of Total Cinema," in which he considered what the "final and complete form" of cinema would be (Bazin 2004, 18). What did cinema want ultimately to become? In this chapter, we will pose that question for AR, and in the next we'll do the same for VR: What are the myths of total AR and VR?

We can trace the myth of total AR at least back to the early 1990s, to what now seems to be the Paleolithic phase of the digital age. David Gelernter published *Mirror Worlds, or the Day Software Puts the Universe in a Shoebox . . . How It Will Happen and What It Will Mean* (1992), forecasting the future integration of our physical and digital worlds: "This book describes an event that will happen someday soon: You will look into a computer screen and see reality. Some part of your world—the town you live in, the company you work for, your school system, the city hospital—will hang there in a sharp color image, abstract but recognizable, moving subtly in a thousand places. This Mirror World you are looking at is fed by a steady rush of new data pouring in through cables" (1).

Even before the internet became the information superhighway in the mid-1990s, and long before the Internet of Things (IoT) in the 2000s, Gelernter understood that the digital era would be characterized by massive flows of data, generated by individuals, organizations, and devices attached to the internet. He wrote that "a Mirror World is some huge institution's moving, true-to-life mirror image trapped inside a computer—where you can see and grasp it whole" (3). At a time when many were predicting with the internet guru John Perry Barlow (1996) that cyberspace would be the new "home of Mind," Gelernter grasped that large amounts of digital data would be concerned with the material conditions of our lives in the physical and social world. What the 1990s called *cyberspace* would become an extension of, not a replacement for, our world.

Gelernter's mirror worlds, however, were generally flat. He was more interested in the flows of data than in their visualization. And when he did discuss visualization, he generally seemed to have in mind 2-D diagrams and dynamic maps. The term *virtual reality* appeared only once in passing (105), and *augmented reality* not at all (hardly surprising, since it had apparently just been coined by Thomas Caudell at Boeing). In the 2010s, ubiquitous smartphones made AR into a medium for delivery of the continuous data flows that Gelernter had in mind. With mobile AR and widely available cellular and Wi-Fi, the pieces were finally in place to realize mirror worlds.

The myth of AR is characterized by two metaphors: the mirror and the cloud. The mirror suggests a digital copy of the world that can be rendered in 2-D and 3-D and overlaid on the visual presentation of your environment (Inbar 2017; Kelly 2019; Nichols 2018). The mirror worlds of AR could display any information that can be associated with locations in our world. This information might be visualized as floating in space above or in the place where it belongs, or it might simply inform the way we interact with the environment. It might be rendered on the screen of a smartphone, tablet, laptop, or desktop. At some time in the future, a headset or glasses that users wear constantly could display the information as they move about in the world, at which point AR would become truly pervasive, a counterpart to the experience of wearing a VR headset and entering a CG metaverse (see chapter 8). The second metaphor, the cloud, uses a familiar term for online data storage. The term *AR cloud* refers specifically to the data describing the physical structure of the world and the 3-D models of things in it, as well as all associated information and data connections residing on internet servers. Applications would download mapping data or 3-D structural views of the mirror world from these cloud servers, and they could also upload information that the user's device gathers and contributes to the general store of data. The mirror is the visualization of the information, and the cloud is the centralized database that contains the information.

Some use the term *AR cloud* to describe not only the servers and their data, but the visualized mirror worlds as well (Inbar 2017). But we will distinguish the AR cloud (or clouds) from the mirror worlds that are generated from the data. The term *mirror world* suggests that AR can reflect aspects of our lived environment by adding layers of information, graphics, video, or audio. The ultimate AR mirror world would be a complete digital doubling of this world. Everywhere we looked—every street, house, office, factory, store—would have its digital twin that our AR devices could make visible. We argue later in this chapter that there will likely never be a single mirror world, just as there will never be a single AR cloud containing all possible relevant data. Instead, there will be partial, filtered views, as well as public and proprietary clouds of data (e.g., those of Google), just as we have throughout the internet today. Google is, in fact, a good place to start. Street View and the data cloud supporting it provide one of the most developed prototypes of a mirror world today.

Google's Prototype of a Mirror World

The illustrations in Gelernter's book were often 2-D maps displaying information in a location, which is exactly what Google Maps has been doing since its launch in 2005 (Wikipedia contributors 2020b). Google Maps interfaces with a geospatial search engine that indexes into a vast and dynamic store of data about restaurants, hotels, theaters, gas stations, traffic conditions, and so on. When Google Street View followed in 2007, two years after Maps, its panoramic views of map locations took the user from two to three dimensions and provided visualization on a scale that deserves to be called a mirror world. The original idea was to map streets from a car with a 360-degree camera mounted on top. Streets in countries all over the (developed) world were soon included (Anguelov et al. 2010), and more recently famous sites (like Zócalo Plaza in Mexico City) and natural features (like the Grand Canyon) have become part of Google Street View. There are now panoramas of the interiors of hundreds of museums and galleries as part of Google Arts and Culture. The views are no longer limited even to the earth: in 2017, the International Space Station was photographed in panoramic format. It is a characteristic Google ambition to add every street, every site, and every interior to its database. Gradually, the company is amassing a photographic reflection of our world. (Other mapping services exist, such as Bing Maps and its StreetSide feature, and Mapillary, but Google dwarfs them.)

Street View is not true VR, because it consists of flat panoramic images rather than 3-D objects drawn in real time. Street View belongs to the long tradition of painted and photographic panoramas (chapter 2) that can be understood as a forerunner of 360-degree video, and users can view Google's mirror world from their laptops, phones, or tablets, as well as in fully immersive VR headsets. If they are using a smartphone and bring up the Street View of their current location, they will see an uncanny double: a photo taken more or less from the point where they are standing, usually from a few months or years past, with differences in construction or street repair and possibly the season of the year.

Street View's database presents the opportunity for AR inside VR. Google Maps now features Live View (figure 6.8), which allows the user to navigate by seeing directions overlaid on the phone's video of the cityscape itself. Because panoramas are images of places in the world, with latitude, longitude, and altitude, they can be augmented with the same geolocated information provided by Google's search features. Google has also integrated Street View panoramas into Google Earth, which uses satellite and aircraft imagery to provide overviews of the planet. As the viewer swoops down and in, the 3-D effect is enhanced (Wikipedia contributors 2020a), and in some cities, there are 3-D models of buildings. Continuing to zoom in takes the user into Street View in that location. Google Earth's VR application allows a fully immersive experience for viewers wearing VR. The integration of satellite imagery, location-based information, 3-D models, and panoramas is more or less seamless.

The Formation of AR Clouds

The Google suite of Maps, Street View, and Earth offers a glimpse of future mirror worlds that will be a far richer combination of physical, commercial, and social data. The amount of location-based data, already vast, is constantly growing and changing. Google's Street View and other services are continually adding to the information of built and natural locations throughout the world. Google's Street View cars are now equipped with light detection and ranging (LIDAR) sensors, as well as cameras to record not only panoramic photos but also depth maps of the cities, towns, and countryside (Simonite 2017).

Although some of this information may be relatively slow to change, none of it is truly permanent. Buildings may endure for decades, although their interiors may be altered and AR clouds may eventually contain floor plans and interior decor. Meanwhile, a vast array of sensing devices in the

Internet of Things, estimated to number 5.81 billion by the end of 2020 (Ranger 2020), is collecting streams of real-time data, much of which is from fixed locations, such as buildings and private homes. These data are likely passing into various company clouds. All the airplanes on the ground and in flight could contribute to the cloud. We can already see thousands of planes in numerous tracking applications such as Flightradar24, on the interactive map on which they congregate over the major airports and fly in trailing lines along the major routes. In the morning, you can watch flocks of intercontinental planes following arcs across the North Atlantic to North America, and in the evening flocks return in similar arcs to Europe. Flightradar24 has an AR mode, a configuration in which you can hold up your phone to view the planes in your area (figure 7.1).

Similar apps offer dynamic traffic maps of other forms of public transportation in major cities, trains, buses, and subways, whose changing locations are sent up to servers.

Smartphone apps and the telcos themselves already track (ostensibly anonymously) the locations of individual users. Forgetting about privacy (as our media culture often seems ready to do), it would be possible to add every individual with a smartphone to the cloud. In the United States

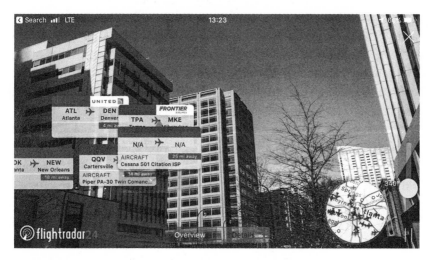

Figure 7.1
The Flightradar24 app gives an AR view of flights as they approach and pass over your location. Reprinted with the permission of Flightradar24.

alone, that was already 81 percent of adults over age eighteen in 2018 (Pew Research Center 2019). Some of these data are collected by tracking features of which most users are unaware— in part because users do not know how or do not care to turn off the location features in their apps, the privacy paradox we'll discuss in chapter 9. As AR apps become common on smart-phones, our trails will be more accurate and complete.

In chapter 4, we described techniques (such as GPS and image tracking) by which an AR device can determine its location more or less precisely. For many purposes, such as many AR games, an application only needs to know its position and orientation relative to its immediate environment— for example, the room. This information might persist only as long as the application is running and never be sent to an online database. But to con-struct a digital mirror of the world that encompasses whole urban environ-ments and landscapes, more than just the relative location of the device is required. What the device sees must be related to its latitude, longitude, and altitude on the surface of the earth. As noted in chapter 4, Google's VPS already connects the local to the global. The phone is constantly tak-ing images of its surroundings and uploading them for comparison; the next logical step (according to the logic of Google's long-term project) is to add the stream of uploaded flat images and eventually 3-D images to the database, to be combined with (eventually) millions of streams from other users to keep the database up to date as streetscapes and landscapes change and to fill in the database where it is incomplete. The technique can be extended to interiors, and Google already has indoor data from museums and other public spaces. Private interiors could be added through tacit or explicit user sharing. In this way, the AR apps in the hands of millions of users do far more than simply calculate the position of the phone itself. They are assembling their own potentially persistent 3-D maps through techniques of computer vision. As sensors such as LIDAR and software algo-rithms such as SLAM determine 3-D structures, scene-understanding algo-rithms can learn to recognize those shapes as tables, chairs, cars, and people (chapter 4).

In this way, the mirror worlds of AR can include increasingly detailed and extensive 3-D replicas. These 3-D models will not necessarily be shown to the users of AR applications. If the user is standing in front of an actual building or in an office with a real desk and chairs, she would not need to see the digital model. Instead of being visualized directly, they will often be invisible presences, used to help anchor the virtual objects in place in the user's field of view. These models can also serve for occlusion, in order to make virtual objects appear to be behind physical ones, adding to the realism. The virtual objects will blend more convincingly into the scene. The invisible presence of physical objects is all the more uncanny precisely because it will go unremarked by most viewers.

The AR cloud need not be limited to 3-D models but can also include text, images, and other data associated with places the models depict. Users of conventional applications are already contributing vast amounts to eventual AR clouds. Users of image services such as Instagram intentionally choose to geotag many of their photos, and about 995 photos are uploaded to Instagram every second (Omnicore 2021b). If even 3 percent of those are geotagged, that makes three million or more a day. All these, along with geotagged tweets, Facebook Messenger posts, YouTube and TikTok videos, and content from other social media applications, do or could enter cloud databases. Combining all these current and future techniques, the collective AR cloud can grow to become an increasingly full digital replica of our physical and built environment, as well as the ephemeral and personal data that accrues around physical locations throughout the world.

There is certainly a great deal of digital information on current servers that is not explicitly geotagged and is beyond the reach of computer vision algorithms: scientific, economic, and demographic data; many social networking pages; tweets, blog posts, essays, and articles; film and television series; digital graphics and art; recorded music; and so on. But even many of these data could be attached metaphorically or descriptively to a location and added to one or another AR cloud. Epidemiological data regarding the spread of disease, such as the coronavirus in 2020, can obviously be connected to various cities and countries. The same is true of economic data, which are meant to characterize some country, region, or people on earth. Fiction and historical documents describe places in the past, present, or (in the case of sci-fi) possible futures. Paintings and graphics often depict a place in the world. The plots of most films and television series unfold in real locations. Most media content is in fact connected with many places rather than just one. The first *Star Wars* movie may take place in a galaxy far, far away, but it was filmed on location here in the Milky Way, and more specifically in Tunisia, Death Valley National Park, and movie studios in the United Kingdom. All films and television shows were filmed or (in the case of CGI) produced somewhere. Tours of such locations have been popular for years: the *Downton Abbey* castle, *The Sopranos'* New Jersey, *Sex and the City's* New York, and so on (Viator 2017).

The mythic total AR Cloud (with a capital C) would contain a plenitude of data. Any real cloud will fall short. It will not include all digital data and will overlook any aspects of our lived experience that cannot be readily converted into digital form, just as internet databases do now. It could nevertheless contain a truly vast amount of digitizable data of many kinds. Each view, each mirror world generated from the AR data store, however, will remain partial for at least two other reasons.

The mythic total AR Cloud (with a capital C) would contain a plenitude of data. Any real cloud will fall short. It will overlook any aspects of our lived experience that cannot be readily converted into digital form.

One AR Cloud?

One reason that every mirror will be a partial view is that it will draw from one or a limited number of cloud servers. The apps and their data will not all be owned by the same company (not even Google), and will not even necessarily be in compatible formats that could be shared. In the last few years, private companies have been developing research and business plans for such clouds. One start-up with the appropriate name 6D.ai (combining a reference to six degrees of freedom with artificial intelligence) has had the supremely ambitious goal of making a cumulative model of the world by crowdsourcing the scanning activities of the users of its app (Matney 2018) and was acquired by Niantic, the maker of *Pokémon Go*. Other start-ups (Scape, Ubiquity6, YouAR, Dent Reality) are pursuing similar crowdsourcing strategies (Fink 2019). Such companies often appear with some fanfare and disappear a year or two later, acquired or out of business, with only a ghost website left behind. Some of those named in the last sentence may well be defunct (like 6d.ai) by the time you read this. Still, the dream of building enormous, crowdsourced databases seems likely to persist. The giant Google, in no danger of disappearing, has the same ambition for its own AR cloud but does not need to rely principally on crowdsourcing. Google's approach can leverage the vast database of images it has collected for Street View to create maps to identify the user's precise location in the world.

All these companies are pursuing the goal of *global localization*. Each use of an AR app gathers data on a particular location, analyzing the buildings and features it can see, and sends these data to servers, where all local data can feed into the global model. For the smaller start-ups, crowdsourcing may be their only source of data; for Google and potentially other giants, such as Facebook or Apple, crowdsourcing would supplement images gathered by their own services or in other ways. The privacy implications are discussed in chapter 9.

Will there ultimately then be one AR Cloud, as the myth of total AR expects, or many smaller ones? The totalizing vision of a single and complete database remains attractive to many technologists working in this field today, such as the founder of 6D.ai, Matt Miesnieks (2018). Ori Inbar, a leading proponent, believes that the single AR mirror world will depend on an aggregate of data, shared among millions and ultimately billions of users. And Kevin Kelly, founding editor of *WIRED*, claims that "someday soon, every place and thing in the real world—every street, lamppost, building, and room—will have its full-size digital twin in the mirrorworld. For now, only tiny patches of the mirrorworld are visible through AR headsets. Piece by piece, these virtual fragments are being stitched together to form a shared, persistent place that will parallel the real world" (Kelly 2019). This is a frank expression of the myth of total AR. If it were to happen, AR would be obeying the inexorable logic of the age of social media.

How complete could a single unified database ever be? Will it in any near future encompass the developing and the developed world? Could it even include all the sites in the zones of coverage, or only certain sites because of legal restrictions or economic interest? (Street View, for example, is more or less banned today in India, Germany, and Austria.) All inhabitants, or only those belonging to certain demographics? There are economic and cultural forces pulling in both directions. Google wants to aggregate and control all digitized knowledge in the world, and that includes the AR cloud. But other companies will certainly offer their own versions. It seems much more likely that we will have a multitude of AR clouds, some as small as a single app, some much more comprehensive, and with them a multitude of limited mirror worlds. Here as elsewhere, the web gives us the potential for an unlimited number of large and small clouds. Ironically, the smaller clouds might be even harder to target for regulation because of their diversity. Companies like Google that control large, centralized clouds of data will be subject to public scrutiny and (at least attempted) legal constraints, although their wealth and size certainly give them the political influence to oppose serious regulation.

Displaying and Interacting with Mirror Worlds

None of the AR clouds may contain the whole world, but the larger ones will still be a plenitude, an almost inexhaustible store of data. And this is the second reason each mirror that reflects that data will be partial. Whether there are dozens of small clouds or a single comprehensive database, the user would never want to view all the information at one time. The data would need to be filtered or it would be overwhelming. A vision of AR gone wild was provided by Keiichi Matsuda's 2016 video *Hyper-Reality*, in which the AR cloud descends upon and literally clouds the vision of its users (figure 7.2). In the video, we see everything from the first-person view of a young woman who wears AR gear as she goes about her job as a personal

Figure 7.2
Hyper-Reality (Matsuda 2016) is a dystopian vision, in which AR takes over our visual environment. Reprinted with permission.

shopper. On the bus, in the street, and in the grocery store, she is deluged with augmented texts and graphics.

Hyper-Reality is a dystopian version of what a totalizing AR cloud might have in store for our media culture. By exaggerating the visual pollution of hundreds of advertisements, reminders, and social media messages competing for our information, the video points out the importance of methods of filtering.

As soon as AR becomes pervasive, with information all around the user and a platform such as a smartphone or headset that is always on, the problem of managing and organizing that information becomes important (e.g., see Grubert et al. 2017). In mirror world applications, information needs to be presented dynamically, appearing in the context in which it is relevant—most obviously, based on where the user is at the moment. Each view needs to reveal or emphasize what is relevant to a particular user at a particular time, just as searches on the web can be filtered (or are filtered automatically by search engines based on our past searches and user profiles). One application or one search might show restaurants in the area; another might show points of cultural interest; another the public tweets and other social media produced by those nearby; and so on. Two users could hold their AR devices up to the same place, and their mirrors would offer different reflections.

In the early days of the web and ubiquitous computing, computer scientists Marc Weiser and John Seeley Brown predicted "the coming age of calm technology." They maintained that properly designed systems could supply our information needs without overwhelming us. A "calm technology . . . engages both the *center* and the *periphery* of our attention, and in fact moves back and forth between the two" (Weiser and Brown 1996). They were not thinking specifically of AR, then in its infancy, but AR mirror worlds would be good candidates for calm design. Interfaces could be fashioned so that the relevant or interesting information (text, images, 3-D models, audio, video) is presented at the center of the user's field of view and attention. More information would wait at the edge or just beyond the user's view or hearing until it is needed. The user could sort the information explicitly, or the system could make decisions for her based on her profile. Instead of Matsuda's scenario in which every icon and text demands immediate attention, like two-year-old children at the store, the icons in a calm design slide patiently past each other, into and out of the center. This ideal depends not only on good design, but on an information infrastructure that is not determined solely by ad-based economics.

It is easy to imagine that the interfaces to the AR clouds will ultimately become extensions of the World Wide Web, which has been the gateway to most digital information since the 1990s, a mirror of our world and yet part of the world at the same time. Immersive web technology (chapter 1) makes it possible for both VR and AR to be rendered visible for us by the standard browsers now used to display web pages. In the case of VR, an immersive web page could become a portal into a totally digital environment—a VR metaverse (see chapter 8). Web AR applications and their clouds would do the opposite, anchoring the web more firmly and visibly in our everyday world.

A Hall of Mirror Worlds

At the end of chapter 6, we suggested that AR applications could be special or everyday. Both kinds draw on an AR cloud for data that they need. Special applications would be tied to events (visiting a museum) or would themselves become an event (playing a game) and so evoke some feeling of novelty. But many applications that draw from an AR cloud, such as navigation and social networking, would fall into the category of the everyday and quickly lose their novelty, especially if the user's AR comes through a future browser and is always on.

Nevertheless, the idea of a mirror world, of a digital reflection of our world, will always have an element of the uncanny. It will be uncanny that the digital copy is constantly updating to reflect our daily world and yet always incomplete. We already have this unsettling feeling when our applications communicate with each other without our conscious intervention—for example, when our photo library organizes photos we have uploaded using metadata or image recognition and suddenly presents our "memories" or when a music service recommends new music based on our previous choices or those of our friends. AR interfaces will reflect these kinds of automatic sorting and categorizing back to us with greater immediacy. The data that we thought we consigned to a particular compartment in our digital storage will return to us as we go through our day. The Internet of Things is also designed to function this way, providing data and regulating our environment without our intervention. This has been touted as a principal virtue of both IoT and its predecessor, ubiquitous computing, as introduced by Mark Weiser decades ago (Weiser 1991). The devices collect data, connect with each other over the internet, and only occasionally need to communicate back to us. Their independence makes them useful and uncanny at the same time.

Multiple AR clouds supporting multiple applications with multiple interfaces and filtered views: taken together, these elements suggest not one mirror world but a hall of mirrors. And like a hall of mirrors at a funhouse, each individual application and filtered view will distort or conceal in order to reveal. It will magnify some aspects of the world it reflects and diminish other aspects or ignore them altogether. If we contrast this vision with the myth of total AR, it becomes apparent that the myth depends on cultural and epistemological as well as technological assumptions. Even if a single, cumulative AR Cloud were technologically feasible, what each mirror world reflects depends on the kinds of human experience and physical information that can be gathered and digitized and on what individuals and the society as a whole value. The fields of information visualization and data studies appreciate this point. As information scholar Yanni Loukissas (2019) points out, "all data are local," dependent on many contexts—including who has the power and authority to decide what qualifies as data. The myth of total AR assumes that there is one established physical and human world that can be reflected transparently in a perfect digital mirror. It does not take account of the uncanny imperfections and gaps in any reflection.

8   The Myth of Total VR: The Metaverse

Bazin's myth of total cinema was the idea that cinema aspired to be "a total and complete representation of reality . . . the reconstruction of a perfect illusion of the outside world in sound, color, and relief" (2004, 19). We are arguing that in effect the corresponding myths of total AR and VR each represent variations or remediations of Bazin's myth for cinema. For AR, the myth of the mirror world envisions the complete digital reconstruction of our lived world. For VR, the myth envisions the perfect 3-D illusion of a world that may or may not resemble the one we live in.

In chapter 7, we dated the myth of total AR back to 1991; the myth of total VR as a fully immersive digital world is decades older. In 1965, Ivan Sutherland imagined the ultimate computer display that would be a looking glass into a "mathematical wonderland," and he imagined the user falling like Alice into such a world (Sutherland [1965] 2009). In 1992, Neal Stephenson published *Snow Crash*, a sci-fi novel that takes place largely in what he called the Metaverse—a virtual space consisting of a single road, one hundred meters wide and over 65,000 km long, spanning the circumference of a virtual planet. Stephenson's Metaverse was a 3-D strip mall, where anyone with the wherewithal could purchase lots and erect virtual buildings. For those who could afford the goggles, the Metaverse became a fully immersive experience, inhabited by millions of avatars—both bots and human users.

When Stephenson was writing in the early 1990s, the two technologies needed to realize his Metaverse were still in development. Although much older as a research infrastructure, the internet was only then finally becoming available through commercial services to the general public. VR was in its first, ultimately abortive, renaissance. There were already some expensive, commercial VR systems, such as Jaron Lanier's VPL headsets, and a few years later, the first game devices appeared. Hollywood, meanwhile, was presenting VR to the public in a dystopian sci-fi film, *The Lawnmower Man* (1992). Hollywood CGI in the 1990s could create compelling 3-D graphics because the graphics were rendered asynchronously, taking minutes or hours for each frame, but most consumer-level computers were not capable of presenting 3-D graphic worlds in real time. When graphics improved in the early 2000s, role-playing games such as *World of Warcraft* (2004) and environments such as *Second Life* (2003) constituted screen-based 3-D metaverses for millions of players. Fully immersive VR is the goal toward which these environments have been heading for decades. In the 2010s, the developer of *Second Life*, Philip Rosedale, promoted his most recent online venture, *High Fidelity*, as a fully VR, decentralized metaverse, although his company eventually scaled back its ambitions to aim at providing businesses with VR conferencing. More recently, social VR environments have emerged, including VRChat, AltspaceVR, and Facebook Horizon.

As with the myth of total AR, the myth of total VR is . . . totalizing. It imagines the metaverse as a single, transparent experience—a perfect, coherent, alternate world. In fact, the metaverses that are emerging take the form of an organizing interface, rather like the desktop metaphor for screen-based computers. This is the metaverse as a world of different VR experiences, each with its own look and feel. The OASIS in *Ready Player One* was imagined as such an encompassing VR environment, filled with different games and environments, from outer space to fantasy kingdoms. But *Ready Player One* was a novel and a movie, not a VR experience. Even in 2018, two decades after *The Lawnmower Man*, its graphics were too elaborate to be generated in real time. So what does a world of VR worlds look like in practice today? It is not as detailed and graphics intensive as the OASIS or the 65,000 km strip mall of *Snow Crash*, but often some kind of large room or house with nature vistas surrounding it (or a space station in a recent Oculus Quest home environment) serves as the staging area for the other VR applications and experiences: the SteamVR home, the Oculus Quest home environments, or the Windows VR Cliff House. In March 2020, just before the launch of *Half-Life: Alyx*, you could incorporate two environments from the game as your SteamVR home. The metaphor of a home suggests a space that belongs to you and that is where you belong. In most of these environments, you can customize your VR home to reflect your tastes, and you may be able to invite people into that home, which transforms it into a modest social space. Social applications such as Hubs start with a metaphor of a larger space for group communication, like a conference room or hall (chapter 6).

The myth of total VR imagines the metaverse as a single, transparent experience—a perfect, coherent, alternate world. In fact, the metaverses that are emerging take the form of an organizing interface, rather like the desktop metaphor for screen-based computers.

Each of these spaces belongs to a different hardware or software platform, and each is a portal into any number of heterogeneous virtual worlds or experiences. These worlds do not have to resemble each other or be consistent in any particular way. One could be a theater, where you can sit and watch movies with friends on a giant screen; another could be a twenty-third-century spaceship heading to Jupiter. Each VR metaverse is in this sense the sum of all the content from all the VR applications that the platform system supports. In the foreseeable future, there will be a relatively small number of metaverses, based on the different thriving platforms. We can compare these metaverse platforms to the operating systems that dominate desktop and laptop computing today: there are only a few important ones (Windows, Mac, and Linux). Just as a few giant tech firms control other aspects of digital media (operating systems, social media, databases), so in VR a few firms are already likely to dominate: Oculus (Facebook), Windows Mixed Reality (Microsoft), Steam (Valve), and perhaps a couple more. We can expect standardization to lead to even fewer platforms in the future, but each platform that ultimately survives will still comprise its own metaverse. It is hard to predict how far the standardization will go and how easy it will be to hop from one metaverse to another. Will it just be a matter of entering a new URL into a headset's browser, or will the user still need a different headset?

Some individual VR gameworlds or social worlds might become so vast that they deserve to be called metaverses on their own. This is the idea behind the OASIS or the original Metaverse in *Snowcrash*. It is odd, by the way, that both of these storyworlds envision a mercilessly capitalist future, but then posit that no other capitalists will manage to maintain their own successful rival metaverses. The notion of a universal VR metaverse is another example of technological universalism: the assumption that one technology will suit all uses—in this case, that everyone will ultimately want the same media experience. The realities of the adoption and creative transformation of media technologies are almost always more nuanced, because different groups, large and small, have different preferences. We can compare the situation with film today, in which some people prefer to go to theaters, while others watch movies on their smartphones.

To qualify as a metaverse, a VR application requires some critical mass of other human participants to interact with. The precursor to the metaverse was a generation of massively multiplayer online role-playing games (MMORPGs) such as *World of Warcraft*, virtual worlds inhabited by thousands or millions of players. In this sense, however, not all virtual worlds are metaverses. The virtual world of an RPG can have a single human inhabitant, one player against many algorithmic, nonplayer characters (NPCs). All the games began that way in the 1990s, before the arrival of online multiplayer functionality, and single-player games are still very popular. Most of the first-generation VR RPGs and shooter games have been solitary experiences as well. Even the highly rated *Half-Life: Alyx* (figure 1.1) is a single-player game. The wasted landscape of City 17 and its zombies and other NPC inhabitants do not constitute a metaverse. VR versions of RPGs will need to engage a community of players who inhabit the world at the same time and interact, and as the base of users with VR headsets grows, such multiplayer games become inevitable.

Conferencing and collaboration for medicine, business, or simply socializing have been promoted as an application area for VR for years (chapter 6). In addition to Facebook Horizon, there are social applications such as Rec Room, a VR space in which to hang out and play casual games. Mozilla is offering Hubs, "an immersive social space delivered through the browser" (White 2018). Such social applications have the potential to become part of a VR metaverse if, as seems likely, they grow in scale to thousands or millions of users and establish a coherent connection to virtual or real locations.

Social Media and the Metaverse

As noted earlier, MMORPGs began to appear in the second half of the 1990s and to thrive in the 2000s. Just as screen-based 3-D games themselves sparked in players' minds the desire for a more immersive experience provided by VR headsets, playing MMORPGs whetted the appetite for social experiences in VR. Players could now dream of participating in quests with fellow players in a fully immersive virtual world. But when Facebook announced its acquisition of the VR start-up Oculus in March 2014, the gaming community reacted with concern and uncertainty (Dredge 2014; Kuchera 2014). Oculus was poised to give gamers the immersive platform that they had been hoping for. And yet Facebook was already a mammoth corporation, and even though the scandals of Cambridge Analytica and fake news were a few years off, the company that made its billions by monetizing its customers' data seemed to have little understanding of gamer culture. The acquisition seemed to be the geeky whim of CEO Mark Zuckerberg. It also seemed to be (and was) yet another expression of the inexorable logic of the tech giants (Google, Apple, and Amazon) to expand into every conceivable application of digital media and information technology.

While the first important applications of inexpensive high-quality headsets would be immersive games, Zuckerberg realized that VR could be a platform for social experiences as well. He wrote: "This is just the start. After games, we're going to make Oculus a platform for many other experiences" (Dredge 2014). These would be shared experiences like going to a basketball game with friends or attending a virtual class with a teacher and other students. By 2019, Zuckerberg was articulating this intuition clearly and including both AR and VR: "What AR and VR do is deliver a sense of 'presence,' where you actually feel like you're there with a person, it's a really deep connection" (Stein and Sherr 2019). Social richness was one of the classic definitions of presence (chapter 5); presence in a metaverse requires being in mediated relationships with others, not just with NPCs.

A VR metaverse can duplicate many of the visual or physical qualities of a cityscape or landscape, but it can also function as an imperfect or even uncanny digital double for face-to-face social interaction, as laptops and mobile devices do now. This social doubling has been a feature of the forerunners of the metaverse for decades, dating back beyond the development of the 3-D MMORPGs to the chat rooms and MUDs of the 1990s. These were online, text-only experiences, in which users conversed with each other and described their actions by typing words on the screen. MUDs served only a fraction of the number of participants of today's *World of Warcraft*, not to mention *Fortnite*. But they fascinated digital cultural critics such as Sherry Turkle because they afforded participants the opportunity to fashion new identities for themselves online (Turkle 1995). Participants in MUDs and chat services created the first digital avatars entirely out of words, often idealized or fantastic versions of themselves. Their characters could change their genders, ages, or other characteristics, which led to a concern about etiquette or honesty. Participants in the chat service of CompuServe, an early online provider, were outraged when an online persona who claimed to be a disabled and disfigured woman and who formed a number of close relationships on that basis turned out to be an able-bodied man (Turkle 1995, 229). The rules for what constituted appropriate role-playing were always being contested. And the later development of 3-D graphic gameworlds changed those rules, or at least user expectations, because adding a visual dimension made the role-playing overt. When you could see a 3-D wizard or a warrior on your screen, you understood that it was not an accurate representation of the person driving that avatar. You did not feel deceived.

As digital media developed and bandwidth expanded in the 2000s, the presentation of self online flourished. Along with 3-D gameworlds, this period saw the arrival of the social media applications and genres that would come to define our current digital media culture: Facebook (2004), YouTube (2005), Twitter (2006), and others. The criticism of social media for compromising our privacy and contributing to our political divisions has done little to stem the growth of social media in general, although users, particularly young users, continue to migrate from one platform to the next. Social media want to extend into, if not appropriate, almost every kind of digital form, and the appropriated forms are eager to cooperate. Every mobile app seems to want to connect to Facebook or offer its own channels for sharing experiences in the form of photos, audio recordings, and videos with friends. The rise of social media encouraged users to present their "real selves." Facebook expects you to appear under your own name, and most users readily take on the ironic role of playing themselves. In the same spirit, YouTube's original motto was "Broadcast yourself." The complicated task of "being yourself" online has been studied by social scientists such as danah boyd (2014) and Zizi Papacharissi (2010), and they have noted how users adopt different personas in different social media apps and for different audiences. However ironic and confusing the results, our society takes the definition of self in social media very seriously and often holds people responsible in a moral and even judicial sense for what they post. People lose their jobs and students get suspended because of Facebook comments or tweets (Mantouvalou 2019). In general, then, while social media have become the places where billions of users are expected to be themselves, role-playing games have flourished as sites where users can be anything but themselves, very often characters in fantasies or science fiction adventures.

VR currently remains, above all, a medium for immersive 3-D games. Gamers constitute the principal market for the headsets from companies such as Oculus, HTC, and Sony, and game companies are rolling out VR versions of popular shooters and role-playing games. But as VR continues to grow, social media will inevitably colonize this medium too, as Facebook is already doing with Oculus. In early 2020, Facebook Horizon, the company's latest proposal for VR as a social medium, became available for the Oculus Quest.

Although its cartoonish graphics and emphasis on games make Facebook's metaverse seem "more like a theme park than a social network" (Stein 2019), the quality of the graphics may not matter for the sense of social presence. The point is to share Facebook's virtual space and communicate through that space with other people as avatars. The graphics, sound, and interactivity just need to be "good enough" to mediate that experience without intruding. Once you can effectively participate on Facebook in your Oculus headset, you will have entered by definition into the largest possible metaverse, with well over two billion inhabitants, and the two forms of online experience (playing a role and playing yourself) will merge.

Among the various current genres of social media, some are more suitable for VR and some for AR. Facebook chose VR, and social networking applications seem better suited to VR, which can create a dedicated space for sharing the "quality time" with friends and acquaintances that Facebook advertises. However, there are features (photo sharing and messaging) in social networking applications that also lend themselves to AR and location-based display, and indeed Facebook claims to want to lead in the development of AR applications as well (Isaac 2017). Short message services like Twitter and photo-sharing services seem better suited to AR, especially as more and more users geotag their tweets and photos, intentionally or by default. One obvious way to experience the AR mirror world will be through a phone app or via AR glasses that display geotagged messages and images as you walk through a building or along a street or a neighborhood. Our immediate experience of the world is in some sense redefined when digital messages and other information can appear around us as we go through our day, when the mirror world becomes visibly present in our world. Video-sharing services such as YouTube and Vimeo seem more suited to VR. Many VR environments already allow their users to sit in a virtual theater and watch traditional or 360-degree videos. Indeed, videos, like photos, can be location-based—especially user-contributed videos that casually record a concert, a speech, an accident, or a police arrest. Such videos might be consumed in an AR application while users are standing or waiting in those locations. Virtual conferencing is possible in both VR (e.g., Hubs) and AR (e.g., Microsoft's Holoportation).

This discussion of social media has apparently led us away from the myth of total VR, and even into AR as an environment for social networking. But in fact, the qualities of immersion and presence associated with VR worlds are characteristic of social media environments as well. Facebook, Instagram, and other social media were always immersive; users can become so engaged that they are said to live in these applications. And we have noted that social presence is one of Lombard and Ditton's (1997) definitions of the phenomenon. The experience of being together in a shared digital environment was what made Facebook so vastly popular on computers and smartphones, and Zuckerberg has grasped that VR can combine social and perceptual presence in new ways for at least a portion of Facebook's community. The myth of total VR fits neatly into his myth of the totalizing power of social media.

The Uncanny Metaverse

While we are witnessing the early formation of metaverses, the development of large, fully functional systems is still years away. For now, the true metaverse has only been realized in science fiction novels and films, and what film versions in particular make clear is that the metaverse will not be a new reality, but rather a new manifestation of VR as a reality medium. Probably the two best-known filmic depictions, *The Matrix* and *Ready Player One*, give us opposing views of what the metaverse(s) could mean for our media culture.

The true metaverse has so far only been realized in science fiction novels and films, and what film versions in particular make clear is that the metaverse will not be a new reality, but rather a new manifestation of VR as a reality medium.

In *The Matrix*, what is left of the human race is living the myth of total VR all the time without realizing it. When Morpheus takes Neo into the Construct, he awakens him to reality by asking him the questions: "What is real? How do you define *real?*" What Neo thinks of as reality is a computer simulation, a metaverse occupied by billions of oblivious human avatars. The Matrix literally redefines reality by interposing itself between the devastated real world and billions of human users living in fetal pods. When Neo takes the red pill and is thrust into the real world, he realizes for the first time in his life that the Matrix is an uncanny double, astonishing in a grotesque way. The film presents the most dystopian view possible of the metaverse, a media experience that people do not willingly enter into and that they cannot escape from.

Ready Player One starts with a similar, if less radical, dichotomy between the economically depressed real world of 2045 and the game world of the OASIS, where millions seek relief from their dreary lives. Unlike Neo, however, the players do remember the real world and spend much of their lives there. In a key departure from Cline's novel, the film ends with a dialogue that parallels the one between Morpheus and Neo in *The Matrix*. Having won the game, the young hero Wade finds himself face to face in the OASIS with a perfectly photorealistic James Halliday, the deceased creator of this gameworld. Wade asks Halliday whether he is an avatar, and Halliday answers no. Then he asks Halliday: "Are you really dead?" When Halliday says yes, Wade continues: "Then what are you?" Halliday doesn't answer. This is the question the film poses about the whole OASIS itself, and it does not want to offer any simple answer. The metaverse of the OASIS is not merely an illusion. The OASIS is real in the sense that it is validated by the experiences of its players; it offers them both perceptual and especially social presence. In fact, the main character Wade first meets his best friends and his girlfriend in the OASIS. The OASIS cannot replace reality, but, like all reality media, it can provide a new perspective on the real. It is no accident that players in the OASIS ultimately return to that other world and try to thrive in both. They are able to appreciate both the suggestive power of total VR and its limits. This is why Spielberg does not ultimately present this VR metaverse as a threat, although, of course, film is at least as important a reality medium as VR for him. Unlike the novel, the film *Ready Player One* suggests how reality media (VR metaverses as well as films) can offer new ways to understand and appreciate our lived world. As we noted in chapter 3, every VR game or experience has the potential to offer new perspectives because it is in some sense an uncanny reflection of our world.

The difference between *The Matrix* and *Ready Player One* is that the former takes the metaverse as myth too seriously. As a replacement for reality, the metaverse will always fall short. By contrast, *Ready Player One* depicts the metaverse as a plenitude of different VR experiences, including but not limited to games, and as yet another realm for social media.

9 Privacy, Public Space, and Reality Media

AR and VR are unique among reality media in the way that they configure the user's relationship to space. Their spaces are not only dynamic but
responsive to the user's movements and, in some cases, active interventions. Vuforia's Chalk app illustrates the power of AR in this respect, but
also suggests the challenge to individual privacy that may be the price the
user pays for that responsiveness:

> When you open Chalk on your iPhone, you are invited to enter a nine-
> digit code that your colleague has sent you. Then the window opens on the
> world around you—your office, say. You are asked to move the phone's view
> around, step back and forth, as the app uses your camera to perform surface
> detection to establish your space. Meanwhile, your colleague is running the
> app and can share your view. Suppose the office printer has jammed. To
> help you fix it, she gives you instructions over the audio channel in the app,
> and at the same time she passes her finger over her screen and creates marks
> with "digital chalk." Her circles or arrows are visible on your screen in real
> time. They also adhere to the surfaces or hang in the air over the printer.

Chalk is an elegant application allowing two people to share the same
augmented view of the world. We can envision using it in the office or at
home—to repair broken devices yourself, to consider how furniture might
be rearranged, to assess fire damage with an insurance agent, to get an estimate of building repair from a contractor, and so on. As you wave the phone
around, it creates a rudimentary digital double of the room for the purposes
of anchoring the chalk marks. In the future, a successor application could
perform more sophisticated scene understanding to distinguish appliances,

computers, machines, people, and everything else the camera sees, and it could send that data to the cloud. PTC, the company that produces Chalk has Terms of Use and Privacy Policy statements that together weigh in at about ten thousand words ("Privacy Policy" n.d.). The terms state that the company may use the data it gathers for any business purpose if the data is not in personally identifiable form. The app could collect information gathered in your apartment or house, as long as it does not identify you personally. The data could then be anonymized by pooling them with similar information from thousands or millions of other users. Companies starting with Google have become remarkably resourceful at finding value by aggregating such data. In this case, if a future version of the application can identify computers, televisions, sound systems, and smart speakers in homes, that information could tell manufacturers and retailers how many of each are in each home and where they are placed. Users of the app could be conducting what amounts to an ongoing survey of consumption patterns for home electronics, which their phones transmit to PTC's cloud silently and for free. We do not know how such data may eventually be used— what kind of digital double will be created and who will have access to it. Our homes that we assumed were private spaces might become public, or at least available to commercial interests, without our explicit knowledge.

We have argued that AR and VR need to be understood in the context of a long history of media technologies. At least since the invention of printing, and particularly since the Industrial Revolution, media have often been used to reconfigure the de facto relationship between public and private and so redefine the social meaning of the two. Photography, the telephone, film, and television all reconfigured public and private. AR and VR, with their emphasis on spatial presentation and manipulation, now join with earlier digital media (especially web services and social media) in another major reconfiguration. In the following pages, we review the historical context in order to better appreciate AR's and VR's impact on public and private, and therefore issues of privacy.

Reconfiguring Public and Private Space

Many of the major media technologies of the nineteenth and twentieth centuries—what Benjamin ([1935] 1968) called *technologies of mechanical reproduction*—have contributed to changing definitions of public and private space through the ways we individually and collectively have chosen to deploy them. The early photographs made by Fox Talbot and Daguerre in the 1840s included streetscapes and landscapes, converting public spaces into private ones that an individual could hang on a wall at home. The reverse was true as well: the interior of homes and other personal spaces could be photographed and shown in public. In the United States, the introduction of the portable and inexpensive Kodak camera by George Eastman in 1888 popularized photography for middle-class consumers, who turned public space into private space by taking pictures of themselves and their families in front of landmarks. They could also take pictures of strangers in public (de Souza e Silva and Frith 2012, 57–58). The camera thus became one of the technologies that sparked legal debate over the right to privacy in the twentieth century.

Around the same time, the telephone was allowing people (at first a wealthy few, but eventually millions) to project their voices across space to contact friends and family far away, extending and modifying the notion of community. Your sense of community was no longer limited to the people in your neighborhood or town (Fischer 1994). The effect was similar to, though perhaps not as profound as, the growth of virtual communities through social media since 2000. Also around the same time, recording technology began to make homes into concert halls and opera houses (Beardsley and Leech-Wilkinson 2020), and a generation later, radio broadcasting conflated public and private space in other ways. A live performance in New York could be heard in your living room, while a fireside chat by President Franklin Roosevelt could feel like a private conversation.

In the reality media of film and television, public and private space was reconfigured yet again. As a dynamic visual medium, film placed audiences in all sorts of fictional and actual spaces. As Benjamin pointed out ([1935] 1968), the film camera penetrated and dissected space, especially through the use of editing that enabled the view to move effortlessly around and through the scene. In film narratives and in newsreels, viewers could glimpse into the private rooms and eavesdrop on the conversations of the wealthy and the famous; they could also take in battlefields in one vista. For the hours that the audience spent in the theater, there was potentially no public or private space that could escape their gaze. With television, all these spaces were brought into the living room (de Souza e Silva and Frith 2012, 71; Williams 2004). What radio did for acoustic space, television did for visual space. Roosevelt's radio broadcasts to the nation in 1933 and 1934 brought his audience listening at home into the White House. Similarly, in 1962, in an event broadcast simultaneously on two of the major television networks, Jacqueline Kennedy transported viewers from their living rooms into the White House for a guided tour. As we have noted, another quality that distinguished television from film was liveness, and television's power to transport the audience in real time was immediately grasped by pioneers such as Edward R. Murrow in his live news program *See It Now* (chapter 5). In the decades that followed, just about every imaginable location on earth became available for broadcast. And not only on earth. By 1969, the world's television audience could view grainy, live video of astronauts walking on the moon.

The internet and the World Wide Web in the 1990s constituted the first wave of digital media's reconfigurations of public and private space. Both the inventors and users of these new media technologies appealed to a spatial metaphor. Users visited websites, which were sometimes thought of as located on servers in various countries, but more often belonged to a new kind of space altogether: *cyberspace*. The term came from William Gibson's cyberpunk novel *Neuromancer* (1984), where it referred to a futuristic database that was also a virtual reality (so a metaverse before Stephenson elaborated the idea). In the 1990s, cyberspace was the practical realization of networked computers throughout the developed world and a metaphor for a technocratic vision of a new public space, what the Grateful Dead lyricist and cyber enthusiast John Perry Barlow (1996) called the "new home of Mind." Barlow articulated a popular expectation that democracy and equality would flourish on the internet because people could exchange information and ideas without gatekeeping institutions and the interference of government. He naively imagined that racial and ethnic prejudices would subside in cyberspace, where everyone was thought to be equal and bodies were largely invisible. (This was still a time when most online communication was conducted through email, bulletin boards, and chat rooms and was limited to text. Images were time-consuming to share, and most users did not have the bandwidth for video.) Just as the political philosopher Jürgen Habermas (1974) argued that newspapers, radio, and television were media of the *public sphere* in the twentieth century, digital optimists argued that the internet could play that role in the twenty-first. As late as 2006, when blogs were already popular but Facebook was just opening up to the general user community, the influential media scholar Henry Jenkins could still conclude *Convergence Culture* with a chapter on digital democracy, claiming that popular culture on the internet could make democracy more participatory (Jenkins 2006).

The notion of *e-democracy* has obviously not fared well in the era of fake news and Russian election hacking. But blogs and social media connected to the internet through desktops and laptops in the 2000s have ended up producing a sphere in which there is no essential difference between public and private speech. Services such as Facebook and Instagram encourage their billions of users to share details, images, and videos of their private life and private spaces with circles of friends, with acquaintances, or with the whole world. As we well know, a tweet that in the past might have been only an ill-considered remark in private can instantly be read by hundreds, thousands, or even millions of followers. YouTube channels make expert lectures and the rants of conspiracy theorists equally available. Digital media give us access to more high-quality information but also more misleading and simply false information than ever before. Democracy is in this sense more participatory than ever before, as Jenkins suggested, but the notion of democratic debate and discussion has been redefined in the process.

When Steve Jobs introduced the iPhone at the Macworld conference in January 2007, he said at first that he was announcing three new devices that day: a phone, an internet browser, and a new iPod. He kept repeating the mantra of three new devices until it became clear to his delighted audience that Apple's new product was in fact one converged device that combined all three capabilities. It was a telephone for personal communication, a browser that connected to the web through cellular service or Wi-Fi, and a media platform for the audio and video that had once been available separately as recordings, film, and television broadcasts.

By 2007, laptops and desktops were already combining two of these three: information from the web and remediations of the photograph, recorded music, cinema, and television. The iPhone completed the bundle by adding the person-to-person communication of the cell phone, putting together in one package all of the most influential technologies of the twentieth century for reconfiguring the public and private. Crucially, the new package could accompany you wherever you went, a key step in the further conflation of public and private space, bringing the 1990s cyberspace together with the physical space of everyday life. We were already used to watching people on the street with a cell phone to their ear. Now it became common for them to hold the phone in front of them, staring at the screen as they walked down the street or stood in line. In the first years of the internet, users visited the digital space on their desktops from the privacy of their home or office. As Wi-Fi spread, they were able to connect their laptops in cafes and libraries, bringing the internet into their spaces. The next step came when millions, eventually billions, of owners of iPhones and Android phones began to take the internet with them wherever they were in public, creating a hybrid space (de Souza e Silva 2006). Now the internet is their more or less constant companion in the private space of the bedroom, indoor public spaces such as cafes and restaurants, cars, trains, busses, the street, and indeed anywhere a cellular or Wi-Fi signal is available.

In one respect, the almost universal use of mobile phones reverses the effect of television. Television brought public space into the viewer's home, and mobile phones and tablets do that as well, displaying streaming video in any personal space. However, mobile technologies also enable the user to appropriate and privatize the public spaces that she visits. When she sits on a park bench absorbed in YouTube videos, especially with headphones to block ambient sound, she renegotiates the relationship between public and private space (Bull 2005, 2015; Campbell 2019; de Souza e Silva and Frith 2012). This process of renegotiation did not begin with mobile phones. Someone could also be sitting on a bench lost in a book, another very portable media technology with the capacity to make public space private. Or they could be listening to music using an MP3 player, fashioning what Michael Bull has called "a privatised auditory bubble" (Bull 2005, 344). Although the mobile phone has largely overtaken specialized portable media technologies like MP3 players, handheld games, or digital reading devices, these still exist and are used by millions. We noted earlier how inexpensive portable cameras, such as Eastman's Kodak, allowed amateurs to appropriate public space for their family photos. Mobile technology extends the impact of earlier media in this sense, which confirms the point Jobs made when he introduced the iPhone as the convergence of earlier media technologies in a single sleek package.

Jenkins (2006) had identified the process of media convergence a year before the iPhone's introduction, and his description captured how the first iPhone and subsequent mobile technologies brought together previously separate media. He was not referring simply to the technologies themselves, but also to their content, the industries that produce the hardware and content, and the practices of consumers. "Convergence involves both a change in the way media is produced and a change in the way media is consumed" (Jenkins 2006, 16). The convergence of various earlier technologies in the smartphone leads to an increasingly thorough interpenetration of public and private space and public and private communication, which also multiples the opportunities for surveillance by government authorities, companies, and individuals.

Surveillance Media

All of the media technologies mentioned thus far have the potential to surveil us. They are all reproductive technologies that can "automatically" record our presence. They can make our private spaces visible to the authorities, and they can change the meaning of a public space for us, where we thought we were anonymous. Among the most important such technologies was CCTV, developed as early as World War II and deployed in a significant way in the 1960s. The UK in particular has a long and enthusiastic history of redefining the nature of public space with surveillance capture for public safety. The London Underground started installing cameras in 1961 (Bradford 2019), and by 2013 there were already 5.9 million CCTV cameras throughout the UK (BBC 2013). The British have even celebrated the achievement with a "National Surveillance Camera Day" ("Surveillance Camera Day: 20 June 2019" 2019). Throughout the developed world, especially in cities, office buildings, and industrial environments, CCTV has made public spaces seem uncanny: we understand that we can be watched not just by passersby, but by the space itself.

Still cameras and video cameras enable one-way surveillance. They surveil the object or person at whom the device is aimed, but not the one doing the listening or watching. Broadcast television and film are also one-way media, but in the other direction; we as the audience get to be the watchers rather than the watched. Yet these media also alter our sense of space through our experience of viewing. The Hollywood or classical style of editing (in film and by extension television) brings us as viewers along with the camera as it moves through the scene, defining what we can see and what we know. The producers of a film or television show do not in general know who is watching it, at least not in real time. As reality media, film and television are dynamic but not interactive. They cannot analyze us as viewers and adapt what we see, as digital media now can.

Digital media have always had the potential to be interactive, but before the internet, this two-way relationship was limited to the user and her computer. In a single-player video game on a stand-alone computer or console, the player responds to what the computer draws on the screen; the computer then responds to the player's response; and so on. All the data remain local. The same is true of a word processor or spreadsheet. But the internet and particularly the web changed the situation; the user's computer became an intermediary between that user and other users or websites on the internet. The advent of commercial applications of the web together with the exponential growth of the internet exposed the population to new forms of surveillance. In 1994, Netscape introduced the *cookie* to store data about the user (Wikipedia contributors 2020i). Throughout the second half of the 1990s, companies were using cookies to get a sense of who their customers were and how to market to them more effectively. As Shoshana Zuboff (2019) argues in *The Age of Surveillance Capitalism*, digital surveillance began in earnest in the early 2000s with the aggressive growth of the advertising system of Google's AdWords (now Google Ads) and AdSense. These systems used the data gathered to match ads with user queries and website visits (Levy 2011). Microsoft's Bing and other services followed suit, constructing a digital double of each of us and making this kind of tracking an unavoidable consequence of our life online. When we use a browser or potentially almost any app on our phones, our changing location may also be tracked. Because our computer or phone can report our online habits and location, even traditional film and television videos may now become surveillance media whenever we view them on any of our digital devices.

The rise of social media in the 2000s added a new dimension to digital surveillance. Facebook's exploitation of its enormous collection of user profiles for targeted advertising was another version of what Google was already doing. The innovation was that users were not simply acquiescing, but voluntarily posting all sorts of personal information online for friends, acquaintances, or the vast general community. Social media is by definition a two-way enterprise: we give and receive information in exchange with our community and in so doing allow companies (and potentially governments) access as well. We participate in our own surveillance. The range of information continues to grow, as users of Facebook, Instagram, and other services upload billions of pictures and videos of themselves and the people and places around them, often now with location data, effectively helping to create a digital copy of the world even without AR.

Social media has revealed the so-called privacy paradox in its starkest form. The paradox is that, when asked, users claim to be concerned about the way apps and services such as those from Google to Facebook compromise their privacy, and yet they seem unwilling to do anything about it (Barth and de Jong 2017; Ooi, Hew, and Lin 2018). They do not stop using Google; they rarely delete their Facebook accounts. In order to install applications, they casually click through end-user licenses, which may run to thousands of words of legalese. In this respect, Vuforia Chalk's agreements are not unusual; in fact, they are more readable than many. Research indicates that current privacy and service policies for online content and mobile applications do not lead to informed users (Obar and Oeldorf-Hirsch 2018). And mobile technology exposes users to new privacy issues that they may not be aware of. Smartphone applications often ask the user, through an initial prompt, to grant blanket agreement to camera access, microphone access, and location services on the device, with little explanation of what kinds of data are gathered (Benford et al. 2015). The danger of location tracking is already apparent in today's phones, through which dedicated companies harvest and sell aggregated location data. These data can reveal not only where an individual goes throughout the day, but often her commercial and social interactions (Valentino-DeVries et al. 2018). If tracking indicates that a phone spends seven hours every weekday at a high school and then nights and weekends at a residential address, it doesn't take much detective work to find out the name of the teacher who owns this phone.

The expanding Internet of Things further extends opportunities for surveillance. IoT is becoming the smartphone's counterpart and complement: you carry the phone from place to place, while many embedded sensors (thermostats, motion detectors, cameras, microphones) stay put and wait for you to come to them. Some people are choosing to wear yet more sensors, such as health and fitness monitors. These sensors may or may not make their presence known to us, but collectively they help to constitute an environment pervaded with devices that look back at us and might be reporting what they find. We may perceive this plenitude of devices as helpful or hostile. Google's Nest thermostat monitors our movements to determine whether we are at home and need to have the heating or cooling adjusted. Amazon's Alexa listens to our requests and provides music and information or turns on other smart devices. But at the same time, these systems could be monitoring us either individually or collectively to commodify our data.

The coronavirus pandemic in 2020 illustrated the ambivalence with which our media culture has come to regard digital surveillance. A number of apps were developed for contact tracing to contain the spread of the virus, and in order to be effective, the apps needed to be provided or sanctioned by governments and very widely used. The German Corona-Warn-App (Robert Koch-Institut 2020) and the Australian COVIDSafe (Australian Government Department of Health 2020) were typical of a decentralized approach to contact tracing in some democracies. They stored the contact information in an encrypted form on the phone, unless and until an individual tested positive for the virus. Then the information was uploaded to a server and used to contact people who had come in close proximity to the tested individual over the previous fourteen days (Germany) or twenty-one days (Australia). Other apps, such as Norway's Smittestop, and the C-19 app (Zoe Global Limited) used in the UK and Sweden stored the data in the cloud. Other, more authoritarian governments created mandatory apps with more direct tracking. It was not long before privacy watchdog groups began to question the security of these systems. Norway, for example, felt compelled to discontinue their app by mid-June 2020 (Singer 2020). The resistance to or acceptance of these apps illustrate the ongoing conflict between the desire for privacy as anonymity on one hand and safety on the other. The question is whether it is more important to be able to pass anonymously through public spaces and come into contact with other people or to view those spaces as safe or welcoming because we believe the danger (in this case of infection) has been reduced. Later stages of the COVID-19 pandemic involved antibody testing to indicate whether a person had had the virus and might therefore be resistant or immune. Either testing serum positive or having a vaccination card confer some sense of security for contacts with others. But that requires making that medical information public or at least available to others. We may come to the point where being immune against the coronavirus or other contagious diseases will be analogous to HIV status among gay men, who voluntarily make their status known to potential partners (Monahan 2020).

In addition to contact tracing, there was an increase in employee monitoring during the pandemic. Automatic monitoring of employees' computer use is nothing new. But when sheltering orders required millions more to work from home, employer interest in tracking software soared. Apps such as Time Doctor, Harvest, Timesheets, and Hubstaff monitor various aspects of employee performance. Hubstaff, for example, can take snapshots of the screen, perform real-time tracking, and determine GPS location (Satariano 2020). Where contact tracing compromises privacy or anonymity in public space, employee tracking invades the private space of the home, the private space of computers at home, or private activities online.

Contact tracing and employee tracking illustrate the ways in which public and private spaces have become increasingly intertwined as new layers of digital technology are introduced and popularized. With mobile applications, the technology was already pushing the process of reconfiguration beyond the internet and into the physical space of everyday life. The advent of both AR and VR furthers and intensifies the trend that digital media have been following since the 1990s.

The Spaces of AR and VR

We have seen how the first phases of digital media gave us a presence in the metaphorical space of the web (cyberspace) and how in the 2010s mobile devices brought more of the physical world under digital surveillance as users came to carry their phones with them throughout the day and imported more of that world into digital space through uploaded photos, videos, and audio. Now the new reality media, and particularly AR on mobile devices, make it increasingly difficult to differentiate between digital and physical space, as well as public and private space. Both AR and VR enable the datafication of space itself. AR systems sense space, VR duplicates space, and both can appear to fill physical space with digital data.

Both AR and VR enable the datafication of space itself. AR systems sense space, VR duplicates space, and both can appear to fill physical space with digital data.

VR in three degrees of freedom (DoF), which tracks only the orientation of your head (chapter 4), may just require you to sit in a chair, and you could presumably do this safely in a cafe or other public place. People would still find this rather strange, but if you do sit in a cafe wearing a VR headset, you will be effectively isolated from the public space around you. Meanwhile, full six DoF VR, which allows you to change position as well as orientation, is generally used in a private physical space at home or in a specially designed space such as a lab, a design studio, or an arcade for multiplayer VR games. Whenever you put on the headset, you leave this physical space (at least partly) behind and enter a created space. This has often been compared to stepping into a film or a game, a visual storyworld that you can inhabit. Unlike the space depicted in film but like that of desktop and console video games, VR space is both dynamic and interactive. VR space is a thorough remediation of filmic space. It is digital and at the same time palpable, in terms of sight, hearing, proprioception, and increasingly touch, as haptic technologies improve. It is not a space that we can live in all the time (not the Matrix), but one that we visit, and the memory of its texture can linger when we take off the headset and rejoin the physical world—pleasantly, or less so in the case of cybersickness.

AR's reconfiguration has a different texture. In the hybrid space of AR, the digital world merges into the physical more palpably and completely than previously with mobile devices and with the Internet of Things alone. AR thus takes our media culture further both in making private space public and in appropriating public space for private use. This reciprocal effect is integral to many applications of AR as a reality medium. Bringing public or commercial data into our home has clear utility—for example, to display furniture that we are thinking of purchasing in our living room or bedroom. By contrast, when we are in a public place and ask our mobile device to display all the data of interest to us (our favorite restaurants or friends' tweets), we are personalizing that space, at least in the moment. We might also leave virtual annotations in the spaces that we visit frequently. Just as we can now save favorite locations in Google Maps on our phones, we could imagine viewing those annotations in Live View, the AR mode in Google Maps (chapter 6).

Such annotations can mark a public space as a place with a personal meaning for each user (Liao and Humphreys 2015). Long before AR, social scientists have been drawing the distinction between space and place (Tuan [1977] 2001), as discussed in chapter 5. The notion of place has no necessary connection to media, but media can help make locations meaningful to us.

Listening to a song or reading a book can become associated with the bench in a park where we were sitting. A photograph can memorialize a location. A book can deepen our understanding of the historical significance of a site we visit, strengthening its aura for us. AR too has the potential to intensify the aura of a historic site for us and therefore give it meaning as a place (chapter 5). And when we use an AR application to annotate a space, we make it into a place whose personal meaning is encoded by the messages we leave there. If the system preserves those messages, we have appropriated that space and made it into "our" place. Because our annotations are virtual, they do not mark the space for anyone else, unless we or the system share them with others. Our appropriations and our meaning do not preclude other appropriations of the same location.

AR might also include others in our acts of meaning making without regard for their wishes. Google's first experiment with a (partial) AR headset, Google Glass, illustrated this danger. The prototype appeared in 2013, and soon a new term emerged for the obnoxious users of the device. Google thus felt compelled to issue guidelines on how not to become a "glasshole" (Osborne 2014). The capacity that people found unsettling was that Google Glass contained a small camera activated by simply touching the frame at your temple (Honan 2013). By 2013, we had already become accustomed to the ubiquitous use of camera phones, but when someone points a phone in your direction, it is obvious what they are doing. Google Glass was essentially a spy cam that could take photos and record videos unobtrusively and upload them effortlessly to a social media account. This appropriation of public space for private use seemed more intrusive than other forms of sanctioned surveillance, especially because a glasshole could take a picture most anywhere, even in the bathroom. In 2015, Google suspended the consumer version of the device, introducing instead an "enterprise" version in 2017, which is still marketed to manufacturers and other businesses, where the spaces are under private control (https://google.com/glass). Glasses or at least sleeker AR headsets continue to be marketed by smaller players such as Vuzix and Snap (Sawh 2020). It seems inevitable that the giants Microsoft (with its HoloLens 2), Apple, and Google will eventually develop affordable glasses for consumers, and the concerns about privacy and the appropriation of public space will return.

Like laptops and desktops, VR and AR devices can record our travels on the web and the data we download and upload. What is new is that VR and

AR devices can also measure and respond to our movements in space. On mobile devices or headsets, they can track the environment through which we are moving, including the people around us. This tracking, inherent in the way AR and VR function to deliver responsive media experiences, means that AR and VR can threaten our privacy to an even greater extent than earlier reality media. When we participate in their mirror worlds and metaverses, AR and VR monitor us at a finer grain than earlier services and applications, conducting a more precise form of Nielsen rating tens or hundreds of times each second. When we gaze at or through VR and AR devices to see a depicted or augmented world, they have an unsettling capacity to gaze back at us.

AR and VR as Surveillance Media

Let's first consider VR's somewhat more limited potential as a surveillance technology. Just as social media, online shopping and banking, and other such services made the transition from laptops and desktops to smartphones and tablets, many will now move into VR. As they transition, commercial services will want to apply surveillance methods established on the 2-D web to tracking the preferences and purchases of players in their metaverses. In addition to the kinds of data that services can collect now (websites visited and time spent, terms searched for, products looked at or purchased, and so on), VR offers the potential for new sources of data through biometric sensing—most immediately eye tracking (Hosfelt 2019), which is in effect gazing at our gaze as we use the system. Eye tracking, not yet standard on today's VR headsets, is the subject of considerable research and development. For the past twenty years, for example, an entire ACM symposium has been held devoted to eye tracking research and applications ("ETRA 2020," n.d.). Both 360-degree video and VR give the viewer a new freedom in relation to traditional film: the viewer can control the camera and do her own editing, deciding where to direct her gaze at any time (chapter 4). The necessary counterpart of that freedom is that the VR system knows where viewers have chosen to gaze and can exploit that knowledge to bind them more tightly into the event loop. Immersion means drawing viewers not only into an imagined world, but also into the algorithm that renders and animates that world, making their actions into data that can be responded to immediately and reported back to the cloud. In a VR shopping mall, for example, the system can report not only which virtual stores users visit, but also which products they look at in the store and along the way. Such experiments are already underway. In 2018, Amazon set up VR kiosks in several shopping malls in India. The kiosks "transport[ed] the shopper into a city filled with Prime Day products—beginning with the fun of a hot air balloon ride. Viewed through an Oculus Rift with full head tracking, the ride lets the shopper briefly see some of the brands and promotions Amazon is featuring before landing in a serene park" (Horwitz 2018). The kiosks did not use eye tracking, but that is clearly a logical step in what Zuboff (2019) calls *surveillance capitalism*.

In VR, the user's virtual location is easy to record because the system has to know where the user is in order to render the virtual world for her. In most current systems, however, her physical location is relatively uninteresting or at least fixed (at home, in a game arcade, in a design studio at work). At some point in the more distant future, VR users might be freer to venture out into the world, if their headsets were equipped with highly accurate and reliable cameras or sensors so that they could see and avoid pedestrians, traffic, and other obstacles (see chapter 10). Then their location in physical space would become as open to surveillance as the location of mobile phone users is today. Until that time, however, AR on mobile devices or headsets raises more serious and diverse privacy concerns than VR.

The capacity of AR to watch us is embodied in the two sets of cameras on your smartphone: one or more looking out at the world and one or more looking back at us. Eye tracking is possible with AR headsets as well (Hosfelt 2019), and the challenges posed by AR eye tracking are greater than with VR because with AR we are out in the world and can be looking at anything or anyone in our environment. AR shopping apps could also ultimately complement location data with eye tracking to determine exactly what products catch the shopper's eye. They could combine that information with data that smartphones or headsets may gather from other sensors to determine a user's gait and, if the user has other wearables such as a smart watch, perhaps heart rate, skin temperature, or other biometric data. All such data could be used for targeting advertisements or for more intrusive purposes.

Pervasive, always-on AR applications have the potential to provide companies or government authorities even more information and with more precision than our current mobile applications do, both by aggregating our habits as consumers and by identifying us as individuals. Ethics researcher Diane Hosfelt describes such a scenario for the familiar example of the furniture app: "Consider the classic AR example: an interior design application that places virtual furniture in your home. I often leave my medications on my nightstand, so that I remember to take them before bed. It's plausible that when I'm redesigning my bedroom, the application will detect the medication, identify it (either by the unique pill shape or by detecting and reading the label), then transmit this information to third parties, which will then use this information to target me for ads related to my condition" (Hosfelt 2019, 3). If the information is sold to pharmacy chains for targeted advertising, it is not far-fetched to imagine that it might eventually become available to prospective employers. A few health insurance companies already use (voluntarily provided) digital information to determine rates (Jeong 2019); they would likely find AR data irresistible if it could be legally gathered.

Pervasive, always-on AR applications have the potential to provide companies or government authorities even more information and with more precision than our current mobile applications do, both by aggregating our habits as consumers and by identifying us as individuals.

As AR makes social media manifest in the visual space around us, it gives familiar threats a new shape, with new opportunities for hacking, trolling, and doxing. Hackers can already use social media to reveal phone numbers and addresses, embarrassing material, and compromising photos of targeted individuals. With AR, they could publish that information in a geolocated tweet so that everyone passing near the individual's home or office would see it. As a medium that interposes itself between the user and her view of the world, AR also multiplies the possibilities for misleading or overwhelming her. Malicious software could tamper with servers by changing or deleting information to be displayed at a location. Keiichi Matsuda (figure 7.2) imagined a future of AR gone wild, in which the user's lived world was hard to see through a forest of floating ads and notices. In that scenario, these data were being provided by companies who at least had some commercial purpose. Hackers could do the same thing with digital signs out of malice or perhaps as a kind of ransomware.

It is, above all, AR's enhanced capacity for location tracking and modeling the world that raises concerns. Current mobile devices track us through cellular data and GPS. The cameras on AR devices that look out at the world can coordinate this data with the images they are taking of buildings and features for surface detection and modeling in order to provide our precise position and at the same time make those models more accurate and complete. AR apps can potentially do this all the time and everywhere, indoors and out, in public buildings and private homes. The technical acronym SLAM (simultaneous localization and mapping) can take on a colloquial and ominous sense when it is the user who gets slammed by the device she is holding. A tech-savvy burglar ring, for example, might gain cloud access to a model of every apartment in her building together with her possessions (everything the camera can see) and also be able to infer whether she is in the apartment at a given time.

Such challenges to privacy depend crucially on who collects and controls the data—in other words, on how global localization is achieved. When we discussed global localization in chapter 7, we noted two current approaches. One is crowdsourcing, in which millions of users' devices contribute to an ever-growing database of 3-D data. The other is a "curated" collection of data maintained by a giant company like Google. Both approaches could result in some company or consortium effectively owning and controlling individuals' data.

This need not be the case. The system could be arranged so that the location-based data that our phone collects are stored privately just for our use, the way we maintain private cloud storage now via services like Dropbox. But that might well require government regulation as commercial AR cloud services would be unlikely to respect our privacy voluntarily when monetizing the data is so valuable. Companies that sell furniture and home appliances, to take one example, would love to know the configuration of bedrooms and other rooms in our houses in order to advertise exactly the right products to us. Ironically, Google's curated approach to its cloud might be easier (at least in theory) to regulate than a wild west of hundreds of small companies relying on crowdsourcing, each with their own data formats and uses.

Some data collected by AR apps could serve the intended purpose without being merged into larger cloud databases at all. An application can create a local mirror world inside a room in the house without GPS and, therefore, without recording the latitude and longitude of its position. It does this by recognizing an image (e.g., a marker on the table for playing a desktop AR game) or by surface detection (chapter 4). For example, an app might construct a model of the surroundings in real time to play an AR game. When the game is over, the model could simply be discarded.

It is also worth noting that there are, at least in theory, public option alternatives to privately owned AR clouds. One such option might be to bypass Google and create an open-source version of Street View. OpenStreetMap (OSM) is an alternative to Google Maps that is seeking to add panoramic images. In the long run, it should be possible to (slowly) enhance something like OSM to include the data needed for global localization. The social and curatorial properties of OSM are already better than Google Street View. New data can be added, and old data deleted or edited, by anyone who cares to do it, allowing problems to be fixed on the ground by motivated parties. The open-source community could create a version of Google's VPS to achieve global localization without its proprietary Street View data.

Virtual Trespassing

One truly novel aspect of AR's reconfiguration of space is that AR enables a kind of virtual trespassing. In 2016, the game *Pokémon Go* introduced millions to the potential of AR gaming (chapter 6). As players' fascination with this locative game mounted, others grew impatient with unexpected crowds of players filling museums, public places, even public transport. PokéStops appeared at cemeteries and memorial sites, among them the Auschwitz-Birkenau State Museum, the National September 11 Memorial & Museum, and the Arlington National Cemetery (Wikipedia contributors 2020h). Amid reports that players were endangering their lives in the pursuit of virtual characters, the Metropolitan Transportation Authority, responsible for the New York City Subway, sent out a tweet on July 11, 2016, imploring users to act responsibly: "Hey #PokemonGO players, we know you gotta catch 'em all, but stay behind that yellow line when in the subway" (NYCT Subway 2016). The Holocaust Museum in Washington, DC, was less patient; its communication director was adamantly opposed to placing so-called PókeStops at their location: "Playing the game is not appropriate in the museum, which is a memorial to the victims of Nazism" (Peterson 2016). The Hiroshima Peace Memorial Park and Auschwitz also sought to remove gamers, who officials felt did not show respect for the sites' historical importance. Niantic, the creator of *Pokémon Go*, was subject to a class-action lawsuit by homeowners for placing virtual objects on their property, a suit that was eventually settled in 2019 (Desatoff 2019).

Marshall McLuhan famously claimed that artists are the early warning system for cultural change, and this has been true for virtual trespassing. As noted in chapter 6, Sander Veenhof and Mark Skwarek created a virtual exhibit in the Museum of Modern Art in New York in 2010, six years before *Pokémon Go*. Their *WeARinMOMA* experience consisted of a number of images and 3-D forms that appeared in a gallery in MoMA when the visitor held up her phone (chapter 6). Another early intervention was *The AR | AD Takeover*, engineered by a group in New York City. They located virtual images on billboards and other public surfaces to "to make evident the consumptive hyper-reality created by commercial advertising." They specifically labelled this installation a "reappropriation of public space" (Biermann et al. 2011, 1). Meanwhile, MoMA remained an appealing target for AR intervention, perhaps because the museum is synonymous with the avant-garde. In 2018, eight artists used an app they called MoMAR to overlay seven paintings in the Jackson Pollack gallery with dynamic imagery and, in one case, a game (DeGuerin 2018). In the fall of that same year, a digital art group named Cuseum "hacked the heist" at the Isabella Stewart Gardner Museum in Boston (discussed in chapter 6). In this case, their purpose was not to add original virtual art but to restore traditional art.

These interventions are not invasions of the user's privacy but instead of the privacy or property rights of the owners of a physical space. They contest the relationship of virtual space to our lived world, raising the questions: Who does virtual space belong to? Can I own the virtual space in and around physical space that belongs (in a legal sense) to me? The control of the virtual space of the web and the internet was and is already highly contested, but prior to the advent of AR, the issue remained in the same realm for digital media as for books, television, film, and radio. It was a question of freedom of expression and its limits in a metaphorical public space. With AR, the metaphor collides with our physical world.

Who can claim the virtual space of public squares, museums, restaurants, or your house? This is obviously a question that could not even have been posed three decades ago. It is a further example of the way in which our lived space and our digital information space (built up since the 1990s) now pervade one another. Everywhere is now digital, and the digital world can now become visible and audible. Through AR's mirror world applications, most every space becomes potentially both public and private.

Uncanny Spaces

VR and AR are among the most recent developments in technologies of surveillance that date back to the video and photographic cameras, as well as audio-recording devices. Placing devices in our environment to watch and listen to us is nothing new; what is novel is the intimate and pervasive relationship that these media technologies set up between users and their space around them. Earlier surveillance technologies were situated somewhere in the world and surveilled us from a vantage point, and even then, the sense that the space was watching us was uncanny enough. Countless spy and detective films feature planted recording devices. Directors of horror and suspense films treat the camera itself as a surveillance device to evoke a sense that the characters are under threat. But the film camera is necessarily only in one location at a time. AR and VR tracking and sensing technologies enlist the whole space to surveil us. They connect us to the space so that we become part of a unified system of mutual sensing and response.

When this works as intended, the space becomes responsive to our actions to achieve the task or create the experience intended. In a VR space, we look in one direction or another, and the system responds by drawing objects in the appropriate perspective. We reach out with our hand or a controller and can pick up virtual objects. In an AR space, the desired information appears in the right place at the right time. But we know that this interaction of tracking and sensing could be transmitted to a server on the internet and recorded. The space around us in that sense is not necessarily benign and cooperative; it may be appropriated to the interests of companies or governments. A rupture occurs when we feel that the space itself has become a surveillance device. Whether benign or threatening, the spaces of AR and VR awaken a sense of the uncanny. It may be a mild surprise at how responsive the space is or a stronger reaction to the threat that the space may pose.

10 The Future

We began this book with the *Arrival of a Train at La Ciotat Station* and argued against the myth that grew up around it. There is another received narrative about film—this one concerning its development as a medium in its first thirty or forty years. That narrative goes like this: The earliest filmmakers borrowed the conventions and practices of other media, particularly live drama. They had their actors perform in front of a stationary camera as if they were onstage. Gradually, filmmakers invented a new language of cinema, particularly by learning to move the camera and edit the film to exploit the unique power of film as a medium. Film was then liberated from its subservience to stage drama and allowed to fulfill its aesthetic destiny as a medium of visual storytelling. The addition of sound (and perhaps color photography) more or less perfected film as a medium.

This idea that film had a single trajectory defined by its inner logic as a medium is as naive as the original La Ciotat myth. *Mainstream* cinema may have reached a kind of formal plateau in the 1930s or the 1950s. But there were always alternative styles, most obviously the achievements of avant-garde filmmakers from the 1920s to the present. Throughout those decades, film has continued to borrow from and remediate stage drama, as well as classic and popular literature and more recently comic books and video games. Film today is deeply embedded in digital media culture—technologically, economically, and in terms of genre.

The myths of total AR and VR make an assumption similar to the received narrative for film: that AR and VR will each develop according to their own inner logic until they achieve perfection. In the case of AR, there will be a single mirror world that perfectly and completely reflects our lived world. In the case of VR, there will be a single metaverse, a benign version of the

Matrix. We have argued throughout this book that the future of both these media will instead resemble the actual history of film since the early twentieth century. AR and VR technologies will certainly improve significantly and may eventually reach a stage of relative stability after which there are only refinements. But their genres will follow trajectories that will converge with and diverge from other genres in our larger media culture. As reality media, AR and VR can continue to develop and to participate in remediating relationships with other media indefinitely.

The Possible, the Plausible, and the Probable

Regarding AR and VR technologies themselves, then, near-term forecasts can be made with some confidence because improvements are almost inevitable. We know that the industry giants (Google, Facebook, Microsoft, and Apple) are continuing to commit some of their seemingly infinite resources to these areas. In five years, the descendants of the Oculus Quest and the HoloLens 2 (whatever they are then called) will perform everything that they can do now, only better. Headsets for VR and glasses for AR will be lighter and more comfortable to wear for long periods. The greater challenge is to predict how large the user communities will be for these increasingly effective devices and what the applications will be.

Enthusiasts for a new technology often assume that what is technically possible must necessarily happen, but we know that this is not so. To take an obvious example, Stanley Kubrick's film *2001: A Space Odyssey* (based on a story by science fiction writer Arthur C. Clarke) was released in 1968 and imagined that in the first year of this century, the United States would have a base on the moon and would be mounting an expedition to Jupiter. From a purely technological perspective, it was not an outlandish idea, considering that NASA was about to land two men on the moon after less than a decade of human space flight. Predictions for lunar colonies and human missions to Mars before the year 2000 were common in the late 1960s (McDougall 2013, xii). If the United States or a group of nations had continued to invest resources in space travel at the levels of the mid-1960s (when NASA's yearly budget was around $40 billion in today's dollars), they might well have a moon base now and be sending a human crew to Mars, if not Jupiter. Likewise, when Cyril Tuschi, whose company makes lightweight AR screens, predicts that AR will entirely eliminate our use of laptop screens, mobile screens, and wearable screens (Peyton 2018), his prediction is not outside the realm of the possible. If everyone were determined to make this happen, Tuschi's vision might come true. But this technological possibility has to reckon with factors such as cost (high-resolution AR systems will be more expensive than screen-based computers for some time), convenience (is it comfortable to write a technical report on a floating virtual screen?), and human inertia (many people won't want to give up their computers and especially their phones).

There is no way to avoid extrapolating existing technologies and trends when making predictions. But such predictions are made difficult by our media culture's creative "misuses" of a technology. In the early days of the mobile phone—say, around 1990—no futurologist would have wanted to predict that short text messaging would remain one of the most important applications thirty years later. Why would anyone send an SMS when they have FaceTime or Skype on their smartphones? Yet in the United States alone, we still send billions of text messages a day (Burke 2018). Meanwhile, the most popular visual communication apps seem to get more and more compressed in time and content. Who could have imagined ten years ago that fifteen-second TikTok videos would become a mass medium? With regard to VR, we can predict that games will likely be the principal application for several years, followed by health and industrial applications and forms of social media. But some new application or remediation of an older form might well emerge in a few years and capture a mass audience: Will AR TikTok experiences be the next thing?

To acknowledge a range of media futures, we appeal to a classification strategy suggested by Anthony Dunne and Fiona Raby (2013) in *Speculative Everything*, borrowed in turn from designer Stewart Candy. They offer a set of terms that can help to distinguish between the technological fantasies and predictions that take into account the conservatism of human users and institutions. Dunne and Raby speak of the possible, the plausible, and the probable—a classification schema they can apply to speculations about all sorts of technological, economic, and social futures. Because we are speculating here about VR and AR, the *possible* comprises every conceivable advance in VR and AR hardware and software: whatever has the remotest chance of occurring, the most utopian (or dystopian) predictions of how the technology will develop and be used. For example, a future in which a large number of people in the developed world choose to spend much of their time in a VR metaverse. The *plausible* includes future developments that seem more likely to happen: for example, the merging of AR into the general interface for mobile phones. A *probable* future is the adoption of VR as a dominant platform for 3-D computer games. The lines between the probable and plausible and between the plausible and possible are obviously blurry, and they become blurrier as we move further into the future. Such speculation has always been the realm of sci-fi novels and films, and it is hard to avoid sounding like science fiction when we try to imagine VR or AR in the coming decades. Novels like *Snow Crash* and films like *Ready Player One* are indeed sources for envisioning the possible (or something just beyond the possible) and, less frequently, the plausible and the probable.

We begin with AR and VR as technological platforms and then consider their ramifying trajectories within our media culture.

The Future of VR Technology

VR is more mature than AR, and the impetus is strong for improving the current technology. These improvements fall on the centerline of the probable because developers know exactly what certain communities (above all, gamers) want: a greater sense of presence and responsiveness through improved headsets and modes of interaction (controllers, hand gestures, and eye and body tracking).

We can predict the general parameters of high-quality yet affordable VR headsets in coming years. In 2020, the Oculus Quest 2 offered a resolution in each of its two eye panels of 1832 by 1920 pixels at consumer prices. There were already headsets costing a few thousand dollars that do considerably better, so it is certain that the price will come down. The same was true of field of view. The Quest had a field of view of about 100 degrees, while the human eyes together have a field of view of about 190 degrees. The expensive StarVR One already beat that with 210-degree horizontal and 130-degree vertical fields of view (Hayden 2020). In five years, CPU and GPU chips will certainly be faster, resulting in more responsive graphics and interactions. The Quest featured hand controllers that had six degrees of freedom (like the headset) and could effectively map hand movements. The Quest also offered a version of hand tracking, so that the user could sometimes dispense with controllers (Robertson 2020). In coming years, VR systems will no doubt be able to track your hands with ever-increasing accuracy. (AR systems such as HoloLens 2 also featured hand tracking and eye tracking, both of which will improve.) The Quest used *inside-out tracking* in general, which means that cameras on the device can create a simple map of the room that you are in. Mapping allowed the Quest's VR scene to show the user the physical obstacles in the room and incorporate them into the game you are playing. In five years, this tracking will improve radically and achieve much better scene understanding. The Quest supported spatial sound at least good enough to differentiate between a noise made on the right or left, in front or beyond. Spatial sound will play a greater role in VR experiences in the coming years. The sense of touch and proprioception will also play a greater role, although it is hard to predict how widely body-suits or omnidirectional treadmills will ever be used.

Although still often portrayed as the "ultimate display," VR will not usurp all the functions of other displays, from huge high-res monitors to phones and smartwatches. Soon, however, its place among accepted media platforms will no longer require justification. VR headsets will no longer seem like toys for geeks, any more than smartwatches do now. VR will, however, continue to construct special realities, apart from the everyday. This will remain a technological necessity as long as VR headsets close us off from the physical world.

VR will continue to construct special realities, apart from the everyday. VR worlds will continue to be metaphoric worlds.

Is there any way that we could imagine VR becoming everyday, like many AR applications? This is in the realm of the possible at some fairly distant point in the future. If tracking and scene understanding become completely reliable, the user could wear a VR headset on the street or in a home or office and still safely navigate the space around her. Rudimentary versions of this technology are appearing, for example, to allow players to wear headsets and play shooter games in a large game space with physical obstacles to hide beyond.

Safe navigation in any arbitrary indoor space or a street is another matter. Right now, some VR headsets, such as the Quest, can display a video view of the outside world to warn a player if she is wandering outside a pre-defined, safe tracking area. But if a future player is going to remain totally in the virtual scene, then that scene would have to be a more or less perfect replica of the street or the office so that when the user comes to a corner or a piece of furniture, she will see in her headset exactly what is in the physical world. The mapping would have to be updated in real time to account for moving objects (such as cars on the street). It would require a miniaturized version of the computer vision systems now being developed (with difficulty) for self-driving cars. Such a system would be VR remediating AR, and it would seem to make more sense just to use see-through AR glasses.

And what about the ultimate vision of perfect VR, the Matrix from the Wachowskis' movie trilogy? As we remember from the films, the Matrix engages all the senses of its human participants or victims so effectively that they cannot distinguish between their physical reality and the simulated world. In the film, the computer system is connected directly to the brains of the participants, who are themselves living in nutritional vats tended by machines. This is obviously far behind the realm of the possible, although it is not unlike the kind of fantasy indulged in by technoenthusiasts like Raymond Kurzweil (2005).

The Future of AR Technology

In 2020, there were already more than three billion smartphones (O'Dea 2020); in other words, almost 40 percent of the world's population carried one around. It is probable that smartphones will remain the most important platform for AR for years to come. Phones have many limitations, but simple AR functions have already been integrated into applications such as Instagram filters, Apple's Animoji feature, and Google Maps Live View. Everyday uses of AR will probably grow in this surreptitious way, by inclusion in other mobile apps, because users will not have to make a decision to acquire any new technology to have access to these features. They may not even realize (or care) that they are using AR when Google Maps provides them with heads-up directions through their mobile screen. But while AR on a smartphone can be casual, a spur of the moment decision to activate a feature, it still has to be intentional. You have to take out your phone and choose the app. A headset or glasses allows AR to become truly pervasive.

AR headsets such as HoloLens 2 and Magic Leap are still for special moments and work activities. Users put them on to play a game or accomplish a design task and then take them off again to go about their day. The question is when and whether, as with conventional glasses, people will put on some future version in the morning and take them off at night. Is it plausible or probable that AR glasses will become ubiquitous? Improvements in style and comfort are inevitable, but to be worn constantly, they will have to be as comfortable and conventional as today's glasses. It is already possible to order prescription lenses for the Magic Leap 1 (Locklear 2018), but the 2020 version of Magic Leap reminds us of the headgear that Cyclops wears to keep his power-beaming eyes from incinerating bystanders.

The technology itself is only one factor; another is the natural tension between peer pressure and an equally natural aversion to change. The pressure to adopt a new technology becomes harder to resist if enough friends and colleagues have already adopted it, which is part of the explanation behind Facebook's success. Everyone is on Facebook because everyone is on Facebook. As more of your friends and coworkers begin to wear AR glasses, you might be more inclined to join them, especially if virtual chatting and conferencing become what Skype and Zoom are today. Sean White, chief R&D officer at Mozilla, was asked, "Do you think AR will someday be pervasive . . . in constant use for us as we go around and do our daily routines?" He replied: "In my mind, this is one of the technologies that is inevitable" (in discussion with the authors, January 3, 2019).

If you are wearing a headset throughout your day, then your access to AR information can be fortuitous, as well as intentional. Information can wait on the periphery and then float into your field of view as needed, based on your location or what you are working on, just as notifications pop up or slide down now on your computer screen or mobile device. Information can also return to the periphery, or indeed just beyond your field of view, waiting for you to turn your head. The prospect of floating information and images suggests that we might become trapped in Keiichi Matsuda's *Hyper-Reality* nightmare (chapter 7). But it is plausible that users will be able to filter the information to suit their preferences, just as they turn notifications off on their mobile devices. What is probable is that at some point in the future (ten years, twenty years?) there will be a portion of the population of the developed world that wears headsets regularly and a portion that does not. The question will be the ratio. Even a massive digital presence such as Facebook is not truly universal. In the United States alone, in 2020, 69 percent of adults used Facebook (Omnicore 2021a), which means that millions of Americans did not. One global estimate for 2020 was 2.6 billion active users (Zephoria 2020), about a third of the world's population. Will

AR headsets be like Facebook in terms of participation, or like the percentage that wears contact lenses, 15 percent or so (GlassesCrafter.com, n.d.)? Projected worldwide, even 15 percent would be hundreds of millions.

The Future Immersive Web

Recall that the acronym XR groups AR and VR (and sometimes other forms of mixed reality, or MR) as *extended realities,* and WebXR refers to the protocol for running XR applications on web browsers, making possible the immersive web (chapter 1). Some of the most popular browsers are committed to supporting WebXR, which means that users will be able to open the browsers on their laptops, smartphones, or tablets and call up VR or AR without having to download a specialized native application. In 2020, Chrome, Edge, and Firefox on Microsoft Windows already supported VR, and Chrome supported AR on Android phones. More browsers and operating systems are expected to follow. Browsers on popular stand-alone headsets such as the Oculus Quest for VR and HoloLens 2 for AR were in various stages of implementation as well. In the near future, then, the immersive web will be available for presenting existing and new genres.

We have seen that VR is mainly for special experiences (games, training simulations, and so on), while AR can be both special (a visit to a museum) and everyday (directions to get you to the museum). We can envision applications in which both reality media are combined in one experience or in which a user moves back and forth between the two or between either one and a screen-based experience, such as reading a webpage. We might call these *mixed* or *casual experiences.* Putting on a VR headset for an hour to take part in a virtual journey to Mars would be a special experience. Reading a web page about the journey, getting a thirty-second VR preview of the journey, and then coming back to the web page would be casual. The *New York Times* has already created several such casual experiences for AR, as noted in chapter 6.

The immersive web is ideal for such casual experiences because it is designed to deliver VR and AR over the ubiquitous medium of the web using the same languages as other web apps. Optimized for speed and graphics on each platform, stand-alone apps will continue to make sense for large, specialized, and lucrative applications, such as AAA games, both in VR and AR, while the immersive web will enable a wide range of developers to create casual and one-time experiences. For thirty years, it has been far easier to create a website than to program a full-blown application for desktop and laptop computers with different operating systems. Organizations and individuals have been able to reach audiences with web apps, even when they cannot afford the cost of making stand-alone apps. The immersive web, realized via WebXR, is poised to open AR and VR design to the same broader class of developers.

The range of possible experiences (special, casual, and everyday) would expand if there were a single headset that could provide both high-quality AR and high-quality VR. We could envision the visitor to a museum wearing a pair of AR-VR glasses while examining an artifact from a pre-Columbian site. The tour application could briefly make the headset opaque and show a panoramic view of the site in Peru where the artifact was found. It would then return to AR mode so that the visitor could continue the tour. But currently VR and AR headsets are constructed differently precisely because AR requires that the user sees the physical world and VR does not. VR headsets achieve wide field of view by taking advantage of the ability to place lenses and screens directly in front of the wearer's eyes, while AR headsets require graphics to be superimposed on an otherwise unimpeded view of the world. Combination headsets will need new optics to provide a wide field of view in AR mode and completely block the view of the world in VR mode. Meanwhile, VR gamers will want headsets with the best resolution and field of view for their immersive games and will continue to buy dedicated VR models. Although it is probable that AR and VR headsets will converge in the future, this convergence might take many years.

Genres of VR and AR

These are the technological affordances and constraints within which VR and AR genres will develop. On analogy with film, the technologies define possibilities but also challenges. Throughout the twentieth century, there was an interdependent relationship between the development of film technologies and the styles, genres, and audiences for film. Everyone knows the impact of the introduction of talkies at the end of the 1920s. Techniques of narrative, acting styles, and camerawork all changed in adapting to the new multimodality. Some actors could not make the shift. Directors and cinematographers had to work under the limitations that sound imposed, especially in the first years, when cameras had to be housed in special booths because of the noise they made. But while this is the most obvious example, film technologies were constantly evolving and film styles were constantly responding and spawning new trajectories. In the 1950s, for example, anamorphic (widescreen) formats encouraged different placement of actors and action. But other formats continued for different film genres, and again experimental filmmakers such as Maya Deren and Stan Brakhage went in quite different directions. More recently, the vogue for 3-D formats in action-adventure movies and animated cartoons affected styles for these genres, while "serious" drama rejected 3-D spectacle. There is an obvious analogy to the development of video games first on screen-based systems and now in VR headsets. As the technology for computer graphics and the speed of CPUs improves, creators of action-adventure and FPS games could strive for greater detail, photorealism, and interactivity, and audiences have come to expect that each new release will improve on the last, while many makers of indie games, "serious" games, or puzzle games continue to work in 2-D or nonphotorealistic styles.

Just as the history of film has never been one of linear improvements in technique and style, there is no reason to think that VR and AR will have such a trajectory either. Like all earlier "new" media, VR and AR enter into chains of remediation with other current media, such as film, television, and screen-based video games. They are already refashioning genres, such as VR video games and 360-degree documentaries, out of familiar ones, such as screen-based shooter games and film documentaries. As the audiences of users and developers in AR and VR expand, their genres and styles will have an impact on the older media genres as well. There are already examples of reverse remediations back into film and television. For example, a sort of *AR effect* can be seen occasionally in films and television shows, where a text message that a character types on a phone floats into the air as if the audience were wearing a headset and could view location-based tweets. AR and VR are in general entering into our media economy through the genres that we identified in chapter 6, and these genres should continue to develop and ramify in the foreseeable future: games, other entertainment and art, cultural heritage, visualizing, learning, training, conferencing, navigation, and social media.

As noted, VR is a more mature technology than AR. Firms that forecast the future of the industry (five or ten years out) seem generally to agree that games and entertainment (such as watching 3-D videos in VR) will continue to dominate (see, e.g., Business Wire 2018; MarketsandMarkets 2019; PR Newswire 2018). We have noted a growing number of successful action-adventure and FPS games. The path is clear for such genres: improvements in computer graphics in headsets and interactivity are what the audience wants, based, for example, on the enthusiastic reception of recent games such as *Half-Life: Alyx*. The development of these VR versions should track closely with that of the desktop versions in the future in terms of themes and visual styles, as well as interaction through more sophisticated AI. A more significant change would be the emergence of massive multiplayer game communities for VR as for the current desktop and console games, such as *Fortnite* and *World of Warcraft*. Massive multiplayer communities seem to be just a matter of time; it simply requires a large enough base of users with headsets. We can also expect that multiplayer VR games will become yet another variation on the Let's Play phenomenon and will be broadcast on Twitch. Other game subgenres should come to thrive in VR as well: for example, physical play and exercise games such as *Beat Saber* or sandbox games such as *Minecraft*. The immersive web will be a good platform for casual VR games, such as puzzles and other games that are played now for a few minutes at a time online in desktop or mobile versions.

It is harder to make predictions about other entertainment and art genres. VR as a platform for watching movies and television with friends may grow into an important niche, when the resolution of VR headsets improves and can offer a viewing experience more like the one that audiences can get on their high-resolution TVs and computer monitors. Videos in 360-degree form have already become an accepted genre in filmmaking, as shown by the film festivals from Sundance to Cannes. Along with the traditional film community, the traditional arts community seems increasingly open to various kinds of interactive VR pieces, as well as other forms of digital art. It is hard to say what styles will develop precisely because the arts community values individual innovation and still has an ambivalent relationship to popular culture, as the examples of VR and AR art in chapter 6 indicate. Contemporary art incorporates, comments on, and critiques popular media forms, and this will continue with VR and AR. All these genres will continue to define niche audiences in comparison to the truly massive audiences for videogames and screen-based film and video, although in the age of digital culture, a niche may encompass thousands or tens of thousands of viewers.

Beyond games and entertainment, the retail and business genres will grow too: particularly health, training, product design, architecture, and construction (Fortune Business Insights 2019, 2020; Marr 2020). VR applications are already in use today in the health industry, as described in chapter 6. But current examples in industrial design, such as Ford and Mercedes using VR to prototype car interiors, often seem as much for publicity (to show how forward looking the car industry is) as actual productivity. Improvements to headsets and hand tracking should lead to greater acceptance. VR simulation and training applications have a significant future, in which VR can be an effective substitute or complement for dangerous or costly on-the-job training (e.g., learning to fly a new commercial airplane or perform a surgical procedure).

The role of social media in VR will surely grow too, as Facebook's purchase and promotion of Oculus suggests. Growing communities of game players and those using VR for entertainment will expect to have social media in their virtual worlds, just as they do on their phones, computers, tablets, and wearables now. We noted how current VR game platforms like Rec Room and SteamVR can easily evolve into platforms for social networking. In general, some of the existing apps and services—not only Facebook and the hugely popular Chinese WeChat—want to add VR to their media empire. It seems likely that as each new social app, such as TikTok, emerges, it will come to have a VR version.

For all the claims that VR is revolutionary, then, this immersive medium fits neatly into many existing generic categories, and all the new versions and applications will have the effect of naturalizing VR within genres that characterize digital media culture today. VR functions as a natural complement to the existing platforms for some genres of games, simulations, training, and most other categories that rely on 3-D computer graphics.

Future applications for AR are perhaps more open for speculation than those for VR. One reason is that AR requires more immediate interaction with the physical world, which expands the scope of applications. Because AR is a less mature technology than VR, it is harder to predict which of the digital genres will be developed and favored by large groups of users. Although games in the tradition of *Pokémon Go* are probably the most profitable applications at present, they have not yet generated the sustained momentum that the lure of immersive gameplay is providing for VR. Similar to games are the advertising applications in magazines that allow the reader to see a 3-D view of what is on the page—for example, a new car. It seems plausible that this genre of AR advertising will continue because these apps are not very costly to produce, even if they remain gimmicks. Along with his VR predictions, Jesse Schell (Ochoa 2015) foresaw the introduction by 2025 of an AR TV companion, an application that works with a television show you are watching. You point your phone's camera at the screen, and it recognizes the image and delivers some information—a *second-screen experience* wherein the second screen is AR on your phone. An early example is augmen.tv, which is already being marketed as video-triggered AR. It also seems probable that there will be more AR applications for maintenance and repair, like the HoloLens HoloGARAGE; such applications were what early AR researchers envisioned in the 1990s.

Two of the most important AR genres in the future may prove to be navigation and again social media. For navigation, Google Maps Live View should prove to be a forerunner of a whole genre of applications to visualize and navigate through AR mirror worlds. And it seems inevitable that social interaction and social media will come to play a defining role in the integration of AR into our media culture. Key social networking services such as Facebook and WeChat have ambitions in AR as well as VR (Xiao 2017). Facebook is apparently developing AR glasses in its Facebook Reality Labs and aiming to deliver them before 2030 (Schomer 2020). In one of his many futuristic predictions, Zuckerberg has suggested that AR will take over the functions of the smartphone, including social networking. AR and VR will then be complementary platforms for extending Facebook's social networking empire.

Mirror Worlds and Metaverses

As we discussed in chapter 7, the location-based data for digital mirror worlds already exist, and those data will vastly increase in the coming years, amassed by the industry giants Google and Facebook and by some smaller companies. They all have an economic incentive to build larger and larger databases for user profiling and ad targeting. For Google, in particular, there is also an ideological motivation because its leaders want to collect and systematize all knowledge in all forms (Foer 2017; Zuboff 2019). Assuming that Google does not succeed in world domination, however, it is not obvious how we avoid a balkanization, in which a few or perhaps several companies each have their own substantial databases.

A single AR mirror world supported by a vast unified database maintained as a public utility might be possible, but multiple data clouds are probable. If there are multiple clouds, will they use a common format and have common rules of access so that all applications can benefit from their data? It seems that a common protocol (set of rules of access) among multiple databases is plausible, if not probable, but that the single cloud will remain a utopian dream in the coming decades. A single, all-encompassing AR Cloud could become a one-stop shopping site for malicious governmental, industrial, and private hacking, making it even harder to maintain privacy than it is with today's social media and online banking, shopping, and bureaucracy. Nevertheless, Sean White (in discussion with the authors, January 3, 2019) is not alone when he says: "I am certain that the AR Cloud will become a reality. My view of the AR Cloud is that there are approaches for us to take in which we can make this a public good and maintain security and privacy." We noted in chapter 7 that enthusiasts such as Kevin Kelly believe this one AR Cloud will be more or less comprehensive: "Eventually, everything will have a digital twin. This is happening faster than you may think" (Kelly 2019).

The myth of total AR, a comprehensive, digital, 3-D replica of the physical world in which every building, street, town, and city is available, will remain on the outer edge of the possible. It seems likely that the 3-D replicas will always be piecemeal and out of date in some areas, just as Google Maps often is today. Parts of the world, presumably in economically advanced countries, will be modeled and available long before others. Different cultures will make different decisions about the advantages and dangers of the cloud. We noted that even Google Street View does not include large parts of Germany and Austria because individuals and the societies as a whole objected. Many countries in Asia and Africa still have very limited coverage.

As a repository of both geolocated data and 3-D models, the cloud will be presented to users as a variety of filtered reflections of our world. It may be visualized in VR, but the more popular and useful applications will be AR, displayed for users in location. Even today's smartphones could bring clouds of AR data to its billions of users. With AR glasses' more sophisticated orientation, recognition, and scanning techniques, the data can be responsive to the user's gaze and movements and connect the cloud more intimately to her immediate surroundings. The metaphor of the cloud is misleading in one sense. It suggests that the digital information is somewhere "up there." But the pervasive display of information through phones and headsets will locate the data all around us. The cloud descends to ground level. Pulling the AR cloud down through a smartphone is highly probable in the coming years. Pulling it down through AR glasses is increasingly plausible.

The metaphor of the cloud suggests that the digital information is somewhere "up there." But the pervasive display of information through phones and headsets locates data all around us.

As for metaverses, we noted in chapter 8 that here too prototypes already exist, such as Rec Room or Hubs, in which participants can take on an avatar and interact with others in shared virtual spaces. It remains unclear how soon these prototypes will engage more fully with the senses of touch and proprioception and whether they will ever converge into a shared metaverse, a VR version of the internet that everyone can access. The myth of total VR is as unlikely as the myth of total AR. Just inside the cone of the possible is something like Stephenson's (1992) Metaverse, a VR space shared by millions, potentially billions of participants. And even Stephenson's Metaverse is not the Matrix. The participants move back and forth between the simulation and the physical world, which means that they know the difference and that a metaverse does not have to create the perfect sensory illusion of reality.

More plausible than the Matrix as metaverse is the one envisioned in *Ready Player One*, a filmic reality of virtual reality. Most Hollywood VR movies are clearly dystopias rather than utopias, but Spielberg's film is unusual because it is a bit of both. All such movies generally confuse audiences about what is possible in VR technology today, but they also fuel popular interest and speculation in the future of VR. The film takes place in 2045; given that time frame, the level of technology depicted, with a few lapses, is on the border between the plausible and the possible. Players can access the OASIS through different devices, some more expensive and full featured than others. The OASIS can offer flawless graphics and sound and haptics, but only when players wear a bodysuit and can move on an omnidirectional treadmill or be suspended on wires. And the OASIS is not the only world the players know; it is an uncanny double of their physical world. Every metaverse in the foreseeable future will be a place that players visit, not live.

As they encompass more and more data, AR mirror worlds will become metaverses too. In one sense, the term *metaverse* should apply even more accurately to AR than VR because AR offers us a meta-universe that combines the digital and physical. All AR digital overlays are annotations of the physical world. AR metaverses will be anchored in our lived world with its physics, its geography, and its social and economic history. They will have the same physical world as their background, and they can at least share the digital data that is publicly available on the internet. They may not share the same 3-D data pertaining to the streetscapes and landscapes of cities and countries. Google already has its own massive proprietary panoramic database in Street View, to which other AR systems might have limited access at best.

Just as for VR, the different AR platforms will mirror the world somewhat differently. Each can offer many of the same experiences, if they have browsers that support the immersive web, but they may also continue to feature their own games and other experiences. It is plausible that there will continue to be separate AR operating systems (for Microsoft, Google, and perhaps Apple), each constituting its own AR metaverse. HoloLens 2, for example, already allows users to call up various applications from a hub, much like Microsoft's VR Cliff House or SteamVR home. Within each metaverse, AR applications may cooperate with one another. AR researchers are already discussing how to allow the user to launch multiple applications at the same time, just as we do now with applications on our laptops and smartphones. In the more distant future, as the technologies for AR and VR move closer together, perhaps ultimately sharing the same headsets and computer vision technologies, the VR and AR metaverses could eventually merge. Something like the Cliff House could become the launching area for both VR and AR experiences. Hybrid worlds will become possible, in which users can move fluidly between AR and VR in the same experience.

Surveillance Capitalism and the Future of Privacy

We discussed in chapter 9 how the new reality media, and especially AR, are leading to a further blurring of the boundaries between public and private space, contributing to the change in our media culture's notion of privacy that social media and commercial tracking of users were already bringing about. Whether AR applications run on smartphones or glasses, they will function as extensions of the other digital media harvesting our data, including stationary computers, laptops, embedded IoT devices, conventional smartphone apps, and wearables. Eye tracking is one refinement that AR and VR add to the data-gathering capacities of these earlier media. As the clouds of location-related data grow, they make ever more data available for surveillance capitalism (Zuboff 2019), as well as traditional government surveillance. Face recognition through all sorts of mobile and stationary camera applications and scene understanding through smartphones and glasses will be ever more significant because they will allow the cloud database to make sense of the image data and compile records of people, objects, and activities associated with places.

What was private is becoming increasingly provisional and negotiable. The value and even the meaning of privacy is now being weighed against the convenience of using online services (such as shopping sites) and location-based services (such as navigation or fitness apps). It is also being weighed against the desire to share texts, images, videos, locations, and online activities (such as listening to streaming music) with friends, acquaintances and the general public. In an appropriately titled book, *It's Complicated* (2014), internet researcher danah boyd argued that teenagers had developed their own nuanced notion of privacy and decorum for using Facebook to share their thoughts and needs with friends. This notion made little sense to their parent's generation, who defined privacy in traditional terms (which ultimately derive from nineteenth-century liberalism). The same will likely be true of the data that AR applications generate and transmit to companies and governments. We trade some "expectation of privacy" with regard to our own personal spaces in return for the ability to navigate and appropriate public space in new ways.

There is also the question of who *we* are. Different cultures and communities have of course always drawn the line between the private and the public in different places, and this remains true in an era of digital media. The Europeans are more concerned about limiting surveillance capitalism than Americans. It is improbable that Germany, for example, will allow AR applications to build commercial or even open-source 3-D databases of buildings and streets, let alone private homes and offices. In

2018, the former CEO of Google, Eric Schmidt, suggested that the internet was breaking into two separate internets, China and the rest of the world (Kolodny 2018). A *New York Times* editorial (Editorial Board 2018) corrected him by suggesting that there will be three: the United States, Europe, and China. While China is erecting its so-called Great Firewall in an effort to keep unwanted sites and messages from reaching its population, Europe has instituted the General Data Protection Regulation to guarantee that its citizens will be able to protect their online privacy. Some policy experts argue for even more internets in what is becoming the "splinternet" (O'Hara and Hall 2018). The main cultural and ethical difference that separates these privacy regimes concerns the roles and prerogatives of governments and businesses. In the United States, businesses have great latitude to store and collect data, and people seem to worry more about government intrusion into their privacy. In Europe, there is more (though by no means absolute) trust in governments; the principal concern is giant corporations such as Google and Facebook. In China, the government is free to use the internet to extend its policy of integrating and monitoring social and cultural life. In addition to censoring what it regards as undesirable materials from within and outside China, the government is gradually integrating digital surveillance technology into its complicated "social credit system" (Matsakis 2019). Face recognition is already widely used. If combined with location tracking and face recognition, AR applications building digital models of China's homes, offices, and public spaces could become a powerful tool for the project of creating an alternative to American and European notions of technological society in the twentieth-first century, which China is seeking to export to trading partners in the developing world.

In addition to the three categories of the possible, plausible, and probable, Dunne and Raby (2013) have a fourth, the *preferable*, which is central to their concept of speculative design. They sketch their fanciful designs to illustrate possible futures (in human transportation, housing, communication, and so on). The goal of these speculations is to get people to consider what futures they would prefer and to stimulate them to work toward those futures. When we speculate about the future of public and private space in AR and VR, the question is not what is possible, because almost any scenario is technologically possible. Societies could heavily regulate the AR clouds and VR metaverses, or simply ban them; they could give corporations free rein to build them and profit from the data; they could adopt the Chinese model and incorporate them into a coherent system of social control. What is preferable depends on belief systems, as well as the politics and economics of individual countries or regions. Thus, it seems very likely that issues of privacy and individual autonomy in the development of AR and VR will divide the internet into at least three zones (the US, Europe, and China) and perhaps more. There will be conflicts within each zone as the societies try to balance competing interests, as well as conflicts between zones, which is inevitable because commerce and communication will at least to some extent cross the barriers between them.

VR and AR Join Media Culture

We can plot our predictions according to the categories provided by Dunne and Raby (2013) by expanding on a diagram that they used in *Speculative Everything* (figure 10.1), which in turn borrowed from the work of Joseph Voros (2003) and Hancock and Bezold (1994).

We can see that those speculations in the outer band and sometimes at the very edge of the possible belong to the myths of total AR and VR. These are the speculations that imagine AR and VR as unique and all-encompassing.

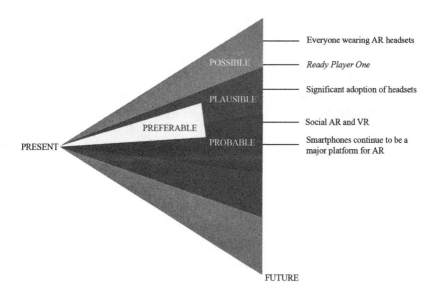

Figure 10.1
The probable, plausible, and possible in VR and AR.

Here AR and VR subsume and replace all other media, if not the lived world itself. They dominate and define our media culture. In the perfect metaverse, or even in the OASIS, all video games, narrative experiences (no need for film), learning experiences (no need for books), visual art, and so on are available in one virtual environment. In the perfect mirror world, screens for watching films and television or reading texts and making (immersive) web sites are all available through our AR glasses (Tuschi's explicit prediction). You just call them up at any size and placement in your field of view. On the other hand, the predictions clustered in the fields of the plausible and probable are those that envision AR and VR as embedded within media culture, participating in remediating relationships and larger genres, in dialogue with forces for cultural and social change. If people continue to use smartphones rather than AR glasses, there will be times when they are not experiencing their everyday world through AR at all. If VR remains a technology for games, entertainment, and certain forms of training, work, and communication, then it continues to compete with other media and options. Not everyone chooses to put on a VR headset to watch a film. It's not just that everyone (one hopes) takes their headsets off to take a walk in the fresh air, but also that some may still prefer to experience film in a theater or on a high-res screen in their living room. Our view is that a multiplicity of platforms and media forms will continue to characterize our media culture, just as it does today.

Although AR and VR will become increasingly imbricated in our media culture, we are not predicting that they will cease to be understood and appreciated as distinct reality media. When Mark Weiser introduced the concept of ubiquitous computing, anticipating the Internet of Things, in the 1990s, he argued that "the most profound technologies are those that disappear. They weave themselves into the fabric of everyday life until they are indistinguishable from it" (1991, 94). Some VR and AR enthusiasts today like to argue something similar: that these technologies will become so natural and integral to our daily lives that we will cease to notice them. They are wrong. However familiar AR and VR become as technologies, they will never become invisible, because media do not function that way. Media always vacillate between transparency and visibility. Sometimes we focus on the content and almost forget the medium, but inevitably the medium reasserts itself. As our engagement with AR and VR becomes widespread, they will become naturalized within various genres in our media culture.

The way VR and AR each construct reality will become familiar. Film ran exactly that course in the twentieth century. It emerged as a highly visible and sometimes controversial medium, and yet its construction of reality eventually became ingrained in our media culture. We saw and to some extent still see the world cinematically; filmic reality has become an aspect of contemporary reality. The same was true of television, and the same will be true of VR and AR, whether their impact is as great as film or television or less. They will constitute part of our media culture, but not disappear into it.

Watching the *Arrival of a Train at La Ciotat Station*, the first audience was astonished to be witnessing photography in motion. The astonishment that early film aroused may have abated in a matter of a few years, but a milder sense of wonder at the way in which film constructed visual reality has never disappeared, not even today in an age dominated by digital media. The double logic of the La Ciotat effect persists. We are aware that we are watching the medium of film, but we are at the same time impressed, no longer astonished, by how uncannily real the moving image seems to be. Our prediction is that this will be true of VR and AR as well. VR may soon be no more or less extraordinary than the experiences of film or television today, and AR may become an everyday media technology. In both AR and VR, the sense of wonder and the uncanny will diminish, but it can always be reawakened by a compelling use—by the ruined city of *Half-Life: Alyx* or by a digital replica of an earthenware horse from a Tang dynasty grave that floats in the air just above a student's desk.

References

4DX. n.d. 4DX home page. Accessed July 21, 2020. http://cj4dx.com/.

ACM SIGRRAPH. n.d. "History of the Organization." ACM SIGGRAPH. Accessed January 6, 2021. https://www.siggraph.org/about/history/.

Acute Art. 2020. "Expanded Holiday—New AR Art by KAWS." Acute Art. https://acute art.com/artist/kaws/.

Alberti, Leon Battista. (1435) 2005. *On Painting*. London: Penguin.

Alberti, Leon Battista. 1804. *Della Pittura e della Statua*, Milano Dalla Società Tipografica de Classici Italiani contrada di S. Margherita, N.° 1118.

Anguelov, Dragomir, Carole Dulong, Daniel Filip, Christian Frueh, Stéphane Lafon, Richard Lyon, Abhijit Ogale, Luc Vincent, and Josh Weaver. 2010. "Google Street View: Capturing the World at Street Level." *Computer* 43, no. 6: 32–38. https://doi.org/10.1109/MC.2010.170.

Arafat, Imtiaz Muhammad, Sharif Mohammad Shahnewaz Ferdous, and John Quarles. 2016. "The Effects of Cybersickness on Persons with Multiple Sclerosis." In *Proceedings of the 22nd ACM Conference on Virtual Reality Software and Technology*, 51–59. Munich: Association for Computing Machinery. https://doi.org/10.1145/2993369.2993383.

Arindam Dey, Mark Billinghurst, Robert W. Lindeman, and J. Edward Swan. 2018. "A Systematic Review of 10 Years of Augmented Reality Usability Studies: 2005 to 2014." *Frontiers in Robotics and AI* 5, no. 37: 1–28. https://doi.org/10.3389/frobt.2018.00037.

Arora, Gabo, and Chris Milk. 2015. *Clouds over Sidra*. 360° film. https://www.with.in/watch/clouds-over-sidra/.

"Augmented Reality Mural Resurrections." n.d. Bowery Wall, New York, NY, 2012. HEAVY. Accessed July 13, 2020. https://www.heavy.io/bowery-info.

Aukstakalnis, Steve. 2016. *Practical Augmented Reality: A Guide to the Technologies, Applications, and Human Factors for AR and VR*. Boston: Addison-Wesley Professional.

Auslander, Philip. 2008. *Liveness: Performance in a Mediatized Culture*. 2nd ed. London: Routledge.

Australian Government Department of Health. 2020. "COVIDSafe App." Australian Government Department of Health. Last updated April 24, 2020. https://www.health.gov.au/resources/apps-and-tools/covidsafe-app.

Bahr, Sarah. 2020. "The Star of This $70 Million Sci-Fi Film Is a Robot." *New York Times*, July 24, 2020, sec. Movies. https://www.nytimes.com/2020/07/24/movies/humanoid-robot-actor.html.

Bailenson, Jeremy N., and Kathryn Y. Segovia. 2010. "Virtual Doppelgangers: Psychological Effects of Avatars Who Ignore Their Owners." In *Online Worlds: Convergence of the Real and the Virtual*, edited by William Sims Bainbridge, 175–186. London: Springer. https://doi.org/10.1007/978-1-84882-825-4_14.

Baldwin, Lily, and Saschka Unseld. 2017. *Through You*. 360° film. https://youtu.be/MQtR4iW3Dqs.

Bancroft, James, Nafees Bin Zafar, Sean Comer, Takashi Kuribayashi, Jonathan Litt, and Thomas Miller. 2019. "Mica: A Photoreal Character for Spatial Computing." In *ACM SIGGRAPH 2019 Talks*, 1–2. Los Angeles: ACM. https://doi.org/10.1145/3306307.3328192.

Barba, Evan, and Blair MacIntyre. 2011. "A Scale Model of Mixed Reality." In *Proceedings of the 8th ACM Conference on Creativity and Cognition*, 117–126. Atlanta, GA: Association for Computing Machinery. https://doi.org/10.1145/2069618.2069640.

Barlow, John Perry. 1996. "A Declaration of the Independence of Cyberspace." Electronic Frontier Foundation. January 20, 2016. https://www.eff.org/cyberspace-independence.

Barrett, Brian. 2018. "The Quiet, Steady Dominance of *Pokémon Go*." *WIRED*. July 6, 2018. https://www.wired.com/story/pokemon-go-quiet-steady-dominance/.

Barth, Susanne, and Menno D. T. de Jong. 2017. "The Privacy Paradox—Investigating Discrepancies between Expressed Privacy Concerns and Actual Online Behavior—A Systematic Literature Review." *Telematics and Informatics* 34, no. 7: 1038–1058. https://doi.org/10.1016/j.tele.2017.04.013.

Baudry, Jean-Louis. 1975. "Le dispositif." *Communications* 23, no. 1: 56–72. https://doi.org/10.3406/comm.1975.1348.

Bazin, Andre. 2004. *What Is Cinema?* Vol. 1. Translated by Hugh Gray. 2nd ed. Berkeley: University of California Press.

BBC. 2013. "5.9 Million CCTV Cameras in UK." *Newsround.* July 11, 2013. https://www.bbc.co.uk/newsround/23279409.

Beardsley, Roger, and Daniel Leech-Wilkinson. n.d. "A Brief History of Recording to ca. 1950." Accessed January 9, 2021. https://charm.rhul.ac.uk/history/p20_4_1.html.

Bekele, Mafkereseb Kassahun, Roberto Pierdicca, Emanuele Frontoni, Eva Savina Malinverni, and James Gain. 2018. "A Survey of Augmented, Virtual, and Mixed Reality for Cultural Heritage." *Journal on Computing and Cultural Heritage* 11, no. 2: art. 7. https://doi.org/10.1145/3145534.

Benford, Steve, Chris Greenhalgh, Bob Anderson, Rachel Jacobs, Mike Golembewski, Marina Jirotka, Bernd Carsten Stahl, et al. 2015. "The Ethical Implications of HCI's Turn to the Cultural." *ACM Transactions on Computer-Human Interaction* 22, no. 5: 1–37. https://doi.org/10.1145/2775107.

Benjamin, Walter. (1935) 1968. "The Work of Art in the Age of Mechanical Reproduction." In *Illuminations*, translated by Harry Zohn, 217–251. New York: Schocken Books.

Bevan, Chris, and David Green. 2018. "A Mediography of Virtual Reality Non-Fiction: Insights and Future Directions." In *Proceedings of the 2018 ACM International Conference on Interactive Experiences for TV and Online Video*, 161–166. New York: Association for Computing Machinery. https://doi.org/10.1145/3210825.3213557.

Bevan, Chris, David Philip Green, Harry Farmer, Mandy Rose, Kirsten Cater, Danaë Stanton Fraser, and Helen Brown. 2019. "Behind the Curtain of the 'Ultimate Empathy Machine': On the Composition of Virtual Reality Nonfiction Experiences." In *Proceedings of the 2019 CHI Conference on Human Factors in Computing Systems*, 1–12. New York: Association for Computing Machinery. https://doi.org/10.1145/3290605.3300736.

Biermann, B. C., Jordan Seller, and Chris Nunes. 2011. "The AR | AD Takeover: Augmented Reality and the Reappropriation of Public Space." Blog. Medium. 2011. https://www.academia.edu/756642/The_AR_AD_Takeover_Augmented_Reality_and_the_Reappropriation_of_Public_Space.

Billinghurst, Mark, Adrian Clark, and Gun Lee. 2015. "A Survey of Augmented Reality." *Foundations and Trends in Human–Computer Interaction* 8, no. 2–3: 73–272. https://doi.org/10.1561/1100000049.

Bizri, Hisham, Andrew Johnson, and Christina Vasilakis. 1998. "Las Meninas in VR: Storytelling and the Illusion in Art." In *Virtual Worlds*, edited by Jean-Claude Heudin, 360–372. Berlin: Springer.

Blast Theory. 2019. "Our Work: Chronology." Accessed July 13, 2020. https://www.blasttheory.co.uk/our-work/.

BMW Group. 2018. "A Glimpse of the Future: BMW Group Uses Virtual Reality to Design Future Production Workstations." BMW Group. Press release. November 20, 2018. https://www.press.bmwgroup.com/global/article/detail/T0287223EN/a -glimpse-of-the-future:-bmw-group-uses-virtual-reality-to-design-future-produc tion-workstations?language=en.

Boeldt, Debra, Elizabeth McMahon, Mimi McFaul, and Walter Greenleaf. 2019. "Using Virtual Reality Exposure Therapy to Enhance Treatment of Anxiety Disorders: Identifying Areas of Clinical Adoption and Potential Obstacles." *Frontiers in Psychiatry* 10. https://doi.org/10.3389/fpsyt.2019.00773.

Bolter, Jay David, and Richard Grusin. 1999. *Remediation: Understanding New Media.* Cambridge, MA: MIT Press.

Bolter, Jay David, Blair MacIntyre, Maribeth Gandy, and Petra Schweitzer. 2006. "New Media and the Permanent Crisis of Aura." *Convergence: The International Journal of Research into New Media Technologies* 12, no. 1: 21–39.

boyd, danah. 2014. *It's Complicated: The Social Life of Networked Teens.* New Haven, CT: Yale University Press.

Bradford, Lowell. 2019. "A History of CCTV Technology: How Video Surveillance Technology Has Evolved." *Surveillance-Video* (blog). August 27, 2019. https://www .surveillance-video.com/blog/a-history-of-cctv-technology-how-video-surveil lance-technology-has-evolved.html/.

Branch, John. 2018. "Augmented Reality: Four of the Best Olympians, as You've Never Seen Them." *New York Times*, February 5, 2018. https://www.nytimes.com/ interactive/2018/02/05/sports/olympics/ar-augmented-reality-olympic-athletes-ul .html.

Bregman, Anrick, Shehani Fernando, and Lucy Hawking. 2017. "*The Party*: A Virtual Experience of Autism—360 Video." *Guardian*, October 7, 2017. https://www .theguardian.com/technology/2017/oct/07/the-party-a-virtual-experience-of-au tism-360-video.

Bull, Michael. 2005. "No Dead Air! The iPod and the Culture of Mobile Listening." *Leisure Studies* 24, no. 4: 343–355. https://doi.org/10.1080/0261436052000330447.

Bull, Michael. 2015. *Sound Moves: iPod Culture and Urban Experience.* New York: Routledge.

Burke, Kenneth. 2018. "How Many Texts Do People Send Every Day (2018)?" *Text Request* (blog). Updated November 2018. https://www.textrequest.com/blog/how -many-texts-people-send-per-day/.

Business Wire. 2018. "Virtual Reality (VR)—Analysis & Outlook on the Global Market to 2026: A $212 Billion Projected Market Opportunity—ResearchAndMarkets.Com."

Business Wire. November 9, 2018. https://www.businesswire.com/news/home/2018
1109005435/en/Virtual-Reality-VR---Analysis-Outlook-Global.

Bye, Kent. 2015. "#245: 50 Years of VR with Tom Furness: The Super Cockpit, Virtual Retinal Display, HIT Lab, & Virtual World Society." *Voices of VR*. Podcast, blog. November 17, 2015. http://voicesofvr.com/245-50-years-of-vr-with-tom-furness-the-super-cockpit-virtual-retinal-display-hit-lab-virtual-world-society/.

Bye, Kent. 2018. "#687: Group Rituals in VR + 2D vs 3D Cinematic Language Experiment." *Voices of VR*. Podcast, blog. August 23, 2018. https://voicesofvr.com/687-group-rituals-in-vr-2d-vs-3d-cinamatic-language-experiment/.

Bye, Kent. 2019. "#847 DocLab: Augmented Audio Tour with Duncan Speakman's 'Only Expansion.'" *Voices of VR*. Podcast, blog. December 12, 2019. https://voicesofvr.com/847-doclab-augmented-audio-tour-with-duncan-speakmans-only-expansion/.

Campbell, Scott W. 2019. "From Frontier to Field: Old and New Theoretical Directions in Mobile Communication Studies." *Communication Theory* 29, no. 1: 46–65. https://doi.org/10.1093/ct/qty021.

Cannes XR. n.d. "Cannes XR." Marché du film Festival de Cannes. Accessed June 9, 2019. https://www.marchedufilm.com/programs/cannes-xr/.

"Canvas_Overview.Pdf." n.d. Accessed January 10, 2021. https://ict.usc.edu/wp-content/uploads/overviews/Canvas_Overview.pdf.

Cartiere, Cameron, and Martin Zebracki. 2015. *The Everyday Practice of Public Art: Art, Space, and Social Inclusion*. New York: Routledge.

Cascone, Sarah. 2018. "Paintings Stolen in America's Biggest Art Heist Have Returned to Their Frames—Thanks to Augmented Reality." Artnet News. March 26, 2018. https://news.artnet.com/art-world/stolen-gardner-paintings-augmented-reality-1252211.

Cauley, Mac. n.d. "The Night Cafe." Mac Cauley. Accessed June 9, 2019. https://www.maccauley.io/portfolio/the-night-cafe/.

Chin, Jimmy, and Ben C. Solomon. 2016. *Man on Spire*. 360° film. New York Times Magazine. https://www.nytimes.com/2016/07/05/magazine/man-on-spire.html.

Cole, Samantha, and Emanuel Maiberg. 2020. "The Endless Battle to Remove Girls Do Porn Videos From Pornhub." *Motherboard. Tech by Vice*, February 6, 2020. https://www.vice.com/en_us/article/9393zp/how-pornhub-moderation-works-girls-do-porn.

Colley, Ashley, Jacob Thebault-Spieker, Allen Yilun Lin, Donald Degraen, Benjamin Fischman, Jonna Häkkilä, Kate Kuehl, et al. 2017. "The Geography of Pokémon GO: Beneficial and Problematic Effects on Places and Movement." In *Proceedings of the*

2017 CHI Conference on Human Factors in Computing Systems, 1179–1192. New York: ACM. https://doi.org/10.1145/3025453.3025495.

Cruz-Neira, Carolina, Daniel J. Sandin, and Thomas A. DeFanti. 1993. "Surround-Screen Projection-Based Virtual Reality: The Design and Implementation of the CAVE." In *Proceedings of the 20th Annual Conference on Computer Graphics and Interactive Techniques (SIGGRAPH '93)*, 135–142. New York: ACM. https://doi.org /10.1145/166117.166134.

Cummings, James J., and Jeremy N. Bailenson. 2016. "How Immersive Is Enough? A Meta-Analysis of the Effect of Immersive Technology on User Presence." *Media Psychology* 19, no. 2: 272–309. https://doi.org/10.1080/15213269.2015.1015740.

Cuseum. 2018. "Hacking the Heist." Accessed January 8, 2021. https://www.hack ingtheheist.com.

Davis, Simon, Keith Nesbitt, and Eugene Nalivaiko. 2014. "A Systematic Review of Cybersickness." In *Proceedings of the 2014 Conference on Interactive Entertainment*, 1–9. Newcastle, NSW, Australia: ACM Press. https://doi.org/10.1145/2677758.2677780.

DeGuerin, Mack. 2018. "Internet Artists Invaded the MoMA with a Guerrilla Augmented Reality Exhibit." Motherboard, Vice. March 5, 2018. https://motherboard .vice.com/en_us/article/8xd3mg/moma-augmented-reality-exhibit-jackson-pollock -were-from-the-internet.

De Guzman, Jaybie A., Kanchana Thilakarathna, and Aruna Seneviratne. 2019. "Security and Privacy Approaches in Mixed Reality: A Literature Survey." *ACM Computing Surveys* 52, no. 6: 1–37. https://doi.org/10.1145/3359626.

de la Peña, Nonny de la. 2015. "The Future of News? Virtual Reality." TEDWomen 2015. May 2015. https://www.ted.com/talks/nonny_de_la_pena_the_future_of_news _virtual_reality#t-266767.

Desatoff, Sam. 2019. "Niantic Settles 2016 Pokemon Go Class Action Trespassing Lawsuit." Yahoo! Finance. September 6, 2019. https://finance.yahoo.com/news/niantic -settles-2016-pokemon-class-194800175.html.

de Souza e Silva, Adriana. 2006. "From Cyber to Hybrid: Mobile Technologies as Interfaces of Hybrid Spaces." *Space and Culture* 9, no. 3: 261–278. https://doi.org/10 .1177/1206331206289022.

de Souza e Silva, Adriana, and Jordan Frith. 2012. *Mobile Interfaces in Public Spaces: Locational Privacy, Control, and Urban Sociability*. New York: Routledge.

DigitalNeohuman. 2010. "Sega VR." YouTube video. April 3, 2010. https://youtu.be /yd98RGxad0U.

Dow, Steven, Manish Mehta, Ellie Harmon, Blair MacIntyre, and Michael Mateas. 2007. "Presence and Engagement in an Interactive Drama." In *Proceedings of the*

SIGCHI Conference on Human Factors in Computing Systems (CHI '07), 1475–1484. https://doi.org/10.1145/1240624.1240847.

Dredge, Stuart. 2014. "Facebook Closes Its $2bn Oculus Rift Acquisition. What Next?" *Guardian*, July 22, 2014. https://www.theguardian.com/technology/2014/jul/22/facebook-oculus-rift-acquisition-virtual-reality.

du Maurier, George. 1878. "Edison's telephonoscope." *Punch Almanac for 1879*, December 9, 1878.

Dunne, Anthony, and Fiona Raby. 2013. *Speculative Everything: Design, Fiction, and Social Dreaming*. Cambridge, MA: MIT Press.

Dupont, Laurent, Marc Pallot, Laure Morel, Olivier Christmann, Vincent Boly, and Simon Richir. 2017. "Exploring Mixed-Methods Instruments for Performance Evaluation of Immersive Collaborative Environments." *International Journal of Virtual Reality* 17, no. 11: 1–29. https://doi.org/10.20870/IJVR.2017.17.2.2888.

Edgerton, Samuel Y. 2009. *The Mirror, the Window, and the Telescope: How Renaissance Linear Perspective Changed Our Vision of the Universe*. Ithaca: Cornell University Press.

Editorial Board. 2018. "There May Soon Be Three Internets. America's Won't Necessarily Be the Best." *New York Times*, October 15, 2018. https://www.nytimes.com/2018/10/15/opinion/internet-google-china-balkanization.html.

"ELITE_SHARP_CTT_Overview.Pdf." n.d. Accessed January 8, 2021. https://ict.usc.edu/wp-content/uploads/overviews/ELITE_SHARP_CTT_Overview.pdf.

Engberg, Maria. 2014. "Polyaesthetic Sights and Sounds: Media Aesthetics in The Fantastic Flying Books of Mr. Morris Lessmore, Upgrade Soul and The Vampyre of Time and Memory." *SoundEffects—An Interdisciplinary Journal of Sound and Sound Experience* 4, no. 1: 21–40. https://doi.org/10.7146/se.v4i1.20370.

Engberg, Maria, and Jay David Bolter. 2020. "The Aesthetics of Reality Media." *Journal of Visual Culture* 19, no. 1: 81–95. https://doi.org/10.1177/1470412920906264.

"ETRA 2020: ACM Symposium on Eye Tracking Research & Applications." n.d. Accessed January 8, 2021. https://etra.acm.org/2020/.

European Commission. 2019. *Communication from the Commission to the European Parliament and the Council*. Leiden, Netherlands: Brill. https://doi.org/10.1163/2210-7975_HRD-4679-0058.

Feiner, Steven. K. 1999. "The Importance of Being Mobile: Some Social Consequences of Wearable Augmented Reality Systems." In *Proceedings 2nd IEEE and ACM International Workshop on Augmented Reality (IWAR'99)*, 145–148. https://doi.org/10.1109/IWAR.1999.803815.

Feinstein, Laura. 2020. "'Beginning of a New Era': How Culture Went Virtual in the Face of Crisis." *Guardian*, April 8, 2020. https://www.theguardian.com/culture/2020 /apr/08/art-virtual-reality-coronavirus-vr.

Fernández del Amo, Iñigo, John Ahmet Erkoyuncu, Rajkumar Roy, and Stephen Wilding. 2018. "Augmented Reality in Maintenance: An Information-Centred Design Framework." *Procedia Manufacturing* 19: 148–155. https://doi.org/10.1016/j .promfg.2018.01.021.

Fink, Charlie. 2019. "Sturfee Begins Large Scale AR Cloud Deployment in Japan, Offers Tools Worldwide." *Forbes*, November 5, 2019. https://www.forbes.com/sites/ charliefink/2019/11/05/sturfee-begins-large-scale-ar-cloud-deployment-in-japan -offers-tools-worldwide/.

Fischer, Claude S. 1994. *America Calling: A Social History of the Telephone to 1940*. Berkeley: University of California Press.

Foer, Franklin. 2017. *World without Mind: The Existential Threat of Big Tech*. New York: Penguin.

Fortune Business Insights. 2019. "Virtual Reality (VR) in Healthcare Market Size, Share & Industry Analysis, By Component (Hardware, Software, and Content), By Application (Pain Management, Education and Training, Surgery, Patient Care Management, Rehabilitation and Therapy Procedures and Post-Traumatic Stress Disorder (PTSD)), and Regional Forecast, 2019–2026." Fortune Business Insights. November 2019. https://www.fortunebusinessinsights.com/industry-reports/virtual-reality -vr-in-healthcare-market-101679.

Fortune Business Insights. 2020. "Virtual Reality Market to Reach USD 120.5 Billion by 2026; Rising Usage in Healthcare & Education Sectors to Aid Growth: Fortune Business Insights™." Globe Newswire. May 15, 2020. http://www.globenewswire.com /news-release/2020/05/15/2034035/0/en/Virtual-Reality-Market-to-Reach-USD-120 -5-Billion-by-2026-Rising-Usage-in-Healthcare-Education-Sectors-to-Aid-Growth -Fortune-Business-Insights.html.

Fox Talbot, William Henry. 1844–1846. *The Pencil of Nature*. London: Longman, Brown, Green and Longmans. http://www.thepencilofnature.com/.

Friedman, Gabe. 2016. "Pokémon Go Invades Auschwitz, US Holocaust Museum and More." *Jewish Telegraphic Agency*, July 12, 2016. https://www.jta.org/2016/07/12 /united-states/pokemon-go-invades-auschwitz-us-holocaust-museum-and-more.

Fritt lopp för de dugliga/Horror over Dalecarlia. 2015. Bombina Bombast. Accessed July 21, 2020. https://www.bombinabombast.com/frittlopp.

Fuscaldo, Donna. 2019. "Half-Life Makes Its Debut as a Virtual Reality Game." *Interesting Engineering*. November 21, 2019. https://interestingengineering.com/half -life-makes-its-debut-as-a-virtual-reality-game.

Fusion. n.d. Fusion LVC home page. Accessed June 7, 2019. https://fusionlvc.com/.

Gatwick Airport Limited. 2017. "Gatwick Installs 2000 Indoor Navigation Beacons Enabling Augmented Reality Wayfinding—A World First for an Airport." Gatwick Media Centre. Press release. May 24, 2017. http://www.mediacentre.gatwickairport .com/press-releases/2017/17_05_25_beacons.aspx.

Gelernter, David Hillel. 1992. *Mirror Worlds: Or the Day Software Puts the Universe in a Shoebox . . . How It Will Happen and What It Will Mean*. New York: Oxford University Press.

Gibson, William. 1984. *Neuromancer*. New York: Ace Books.

GlassesCrafter.com. n.d. "What Percentage of the Population Wears Glasses?" Accessed March 4, 2019. http://glassescrafter.co.uk/information/percentage-popula tion-wears-glasses.html.

Google LLC. *Google Arts & Culture*. V. 8.0.54. Google LLC, 2016. https://apps.apple .com/app/arts-culture/id1050970557.

Google Developers. 2019. "Google Keynote (Google I/O'19)." YouTube video. May 7, 2019. https://youtu.be/lyRPyRKHO8M.

Gorky, Maxim. (1896) n.d. "Life Devoid of Words." *Lapham's Quarterly*. 1896. https://www.laphamsquarterly.org/arts-letters/life-devoid-words.

Grau, Oliver. 2003. *Virtual Art: From Illusion to Immersion*. Cambridge, MA: MIT Press.

Greengard, Samuel. 2019. *Virtual Reality*. Cambridge, MA: MIT Press.

Grubert, Jens, Tobias Langlotz, Stefanie Zollmann, and Holger Regenbrecht. 2017. "Towards Pervasive Augmented Reality: Context-Awareness in Augmented Reality." *IEEE Transactions on Visualization and Computer Graphics* 23, no. 6: 1706–1724. https://doi.org/10.1109/TVCG.2016.2543720.

Gunning, Tom. 1986. "The Cinema of Attraction: Early Film, Its Spectator and the Avant-Garde." *Wide Angle* 8, no. 3–4: 63–70.

Gunning, Tom. 2009. "An Aesthetic of Astonishment: Early Film and the (In)credulous Spectator." In *Film Theory and Criticism*, edited by Leo Braudy and Marshall Cohen, 818–832. New York: Oxford University Press.

Habermas, Jurgen. 1974. "The Public Sphere: An Encyclopedia Article (1964)." *New German Critique*, no. 3 (Autumn): 49–55.

Hancock, T., and C. Bezold. 1994. "Possible Futures, Preferable Futures." *Healthcare Forum Journal* 37, no. 2: 23–29.

Hayden, Scott. 2019. "'High Fidelity' Refocuses on Enterprise Market, Lays off 25% of Staff." *Road to VR*, May 8, 2019. https://www.roadtovr.com/high-fidelity-refocuses -enterprise-market-lays-off-25-staff/.

Hayden, Scott. 2020. "Ultra-Wide FOV Headset StarVR One Priced at $3,200, Selling to Enterprise Only." *Road to VR*, May 4, 2020. https://www.roadtovr.com/starvr-one -launch-acer-starbreeze/.

Heater, Brian. 2020. "Pornhub Removes All Unverified Content, Following Reports of Exploitation." *Techcrunch*, December 14, 2020. https://techcrunch.com/2020/12/14 /pornhub-removes-all-unverified-content-following-reports-of-exploitation/.

Herrera, Fernanda, Jeremy Bailenson, Erika Weisz, Elise Ogle, and Jamil Zaki. 2018. "Building Long-Term Empathy: A Large-Scale Comparison of Traditional and Virtual Reality Perspective-Taking." *PLOS ONE* 13, no. 10: e0204494. https://doi.org /10.1371/journal.pone.0204494.

Honan, Mat. 2013. "I, Glasshole: My Year with Google Glass." *WIRED*, December 30, 2013. https://www.wired.com/2013/12/glasshole/.

Horwitz, Jeremy. 2018. "Watch Amazon's VR Kiosks Transform the Future of Shopping." VentureBeat. July 12, 2018. https://venturebeat.com/2018/07/12/watch-ama zons-vr-kiosks-transform-the-future-of-shopping/.

Hosfelt, Diane. 2019. *Making Ethical Decisions for the Immersive Web*. ArXiv paper 1905.06995. May 14, 2019. http://arxiv.org/abs/1905.06995.

Huang, Jensen. 2020. "NVIDIA GTC May 2020 Keynote Pt2: NVIDIA RTX - A New Era for Computer Graphics." YouTube video. May 14, 2020. https://www.youtube .com/watch?v=BeScfkCm3b4.

Hughes, John F., Andries van Dam, James D. Foley, Morgan McGuire, Steven K. Feiner, and David F. Sklar. 2014. *Computer Graphics: Principles and Practice*. 3rd ed. Upper Saddle River, NJ: Addison-Wesley.

Inbar, Ori. 2017. "ARKit and ARCore Will Not Usher Massive Adoption of Mobile AR: The AR Cloud Will." *Super Ventures Blog* (blog). September 12, 2017. https:// medium.com/super-ventures-blog/arkit-and-arcore-will-not-usher-massive-adop tion-of-mobile-ar-da3d87f7e5ad.

Isaac, Mike. 2017. "Mark Zuckerberg Sees Augmented Reality Ecosystem in Facebook." *New York Times*, April 18, 2017. https://www.nytimes.com/2017/04/18/tech nology/mark-zuckerberg-sees-augmented-reality-ecosystem-in-facebook.html.

Jenkins, Henry. 2006. *Convergence Culture: Where Old and New Media Collide*. New York: New York University Press.

Jeong, Sarah. 2019. "Insurers Want to Know How Many Steps You Took Today." *New York Times*, April 10, 2019. https://www.nytimes.com/2019/04/10/opinion/insurance -ai.html.

Jerald, Jason. 2016. *The VR Book: Human-Centered Design for Virtual Reality*. New York: Association for Computing Machinery and Morgan & Claypool Publishers.

Johnson, Charles S., Jr. 2017. *Science for the Curious Photographer: An Introduction to the Science of Photography*. New York: Taylor & Francis.

Jud, Lukas, Javad Fotouhi, Octavian Andronic, Alexander Aichmair, Greg Osgood, Nassir Navab, and Mazda Farshad. 2020. "Applicability of Augmented Reality in Orthopedic Surgery—A Systematic Review." *BMC Musculoskeletal Disorders* 21, no. 1: 103. https://doi.org/10.1186/s12891-020-3110-2.

Jung, Timothy, M. Claudia tom Dieck, Hyunae Lee, and Namho Chung. 2016. "Effects of Virtual Reality and Augmented Reality on Visitor Experiences in Museum." In *Information and Communication Technologies in Tourism 2016*, edited by Alessandro Inversini and Roland Schegg, 621–635. Cham: Springer International Publishing.

Katz, Miranda. 2018. "Augmented Reality Is Transforming Museums." *WIRED*, April 23, 2018. https://www.wired.com/story/augmented-reality-art-museums/.

Keene, Sam. 2018. *Google Daydream VR Cookbook: Building Games and Apps with Google Daydream and Unity*. Boston: Addison-Wesley. https://www.informit.com/store/google-daydream-vr-cookbook-building-games-and-apps-9780134845517.

Keijzer, Hugo, dir. 2016. *The Invisible Man*. 360° VR short film. Midnight Pictures. https://vimeo.com/207495856.

Kelly, Kevin. 2019. "AR Will Spark the Next Big Tech Platform—Call It Mirrorworld." *WIRED*, February 12, 2019. https://www.wired.com/story/mirrorworld-ar-next-big-tech-platform/.

Kolodny, Lora. 2018. "Former Google CEO Predicts the Internet Will Split in Two—and One Part Will Be Led by China." CNBC. September 20, 2018. https://www.cnbc.com/2018/09/20/eric-schmidt-ex-google-ceo-predicts-internet-split-china.html.

Kristof, Nicholas. 2020. "Opinion | The Children of Pornhub." *New York Times*, December 4, 2020. https://www.nytimes.com/2020/12/04/opinion/sunday/pornhub-rape-trafficking.html.

Kuchera, Ben. 2014. "VR Game Creators React to Facebook's $2 Billion Purchase of Oculus." Polygon. March 25, 2014. https://www.polygon.com/2014/3/25/5547584/facebook-buys-oculus-rift-game-developers-reaction.

Kurzweil, Ray. 2005. *The Singularity Is Near: When Humans Transcend Biology*. New York: Viking Press.

Lanier, Jaron. 2017. *Dawn of the New Everything: Encounters with Reality and Virtual Reality*. New York: Henry Holt and Company.

Le Forum des images. 2017. "Gabo Arora: How Can VR Promote Social Causes?" YouTube video. June 30, 2017. https://youtu.be/E91qyR0J8FQ.

Lee, Kangdon. 2012. "Augmented Reality in Education and Training." *TechTrends* 56, no. 2: 13–21. https://doi.org/10.1007/s11528-012-0559-3.

Levere, Jane L. 2019. "Meeting Old Masters, Rowing with Vikings—in Augmented Reality." *New York Times*, March 12, 2019. https://www.nytimes.com/2019/03/12/arts /augmented-reality-app-exhibits.html?smid=nytcore-ios-share.

Levy, Richard M. 2012. "The Virtual Reality Revolution: The Vision and the Reality." In *Virtual Reality—Human Computer Interaction*, edited by Tang Xinxing, 21–40. London: IntechOpen. https://doi.org/10.5772/51823.

Levy, Steven. 2011. *In the Plex: How Google Thinks, Works, and Shapes Our Lives*. New York: Simon & Schuster.

Liao, Tony, and Lee Humphreys. 2015. "Layar-ed Places: Using Mobile Augmented Reality to Tactically Reengage, Reproduce, and Reappropriate Public Space." *New Media & Society* 17, no. 9: 1418–1435. https://doi.org/10.1177/1461444814527734.

Liestøl, Gunnar. n.d. "Situated Simulations Lab." SitsimLab. Accessed June 6, 2019. http://sitsim.no/.

Locklear, Mallory. 2018. "Magic Leap Prescription Lenses Are Available Now for $249." *Engadget*, December 19, 2018. https://www.engadget.com/2018-12-19-magic -leap-prescription-lenses-available-249.html.

Loeffler, John. 2019. "The History and Science of Virtual Reality Headsets." Interesting Engineering. February 28, 2019. https://interestingengineering.com/the-history -and-science-of-virtual-reality-headsets.

Lombard, Matthew, and Theresa Ditton. 1997. "At the Heart of It All: The Concept of Presence." *Journal of Computer Mediated Communication* 3 (2). https://onlinelibrary .wiley.com/doi/full/10.1111/j.1083-6101.1997.tb00072.x.

Loukissas, Yanni Alexander. 2019. *All Data Are Local: Thinking Critically in a Data-Driven Society*. Cambridge, MA: MIT Press.

MacIntyre, B. 2020. "Remote Conference Participation in Social Virtual Worlds." *Semantic Scholar*. https://www.semanticscholar.org/paper/Remote-Conference-Partici pation-in-Social-Virtual-MacIntyre/05450173ca956988a6f9f84484db9c561e61d054.

MacIntyre, Blair, and Trevor Smith. 2018. "Thoughts on the Future of WebXR and the Immersive Web." In *2018 IEEE International Symposium on Mixed and Augmented Reality Adjunct (ISMAR-Adjunct)*, 338–342. Munich: IEEE. https://doi.org/10.1109 /ISMAR-Adjunct.2018.00099.

Mantouvalou, Virginia. 2019. "'I Lost My Job over a Facebook Post: Was That Fair?' Discipline and Dismissal for Social Media Activity." *International Journal of Comparative Labour Law and Industrial Relations* 35, no. 1: 101–125.

MarketsandMarkets. 2019. "Virtual Reality Market with COVID-19 Impact Analysis by Offering (Hardware and Software), Technology, Device Type (Head-Mounted Display, Gesture-Tracking Device), Application (Consumer, Commercial, Enterprise, Healthcare) and Geography—Global Forecast to 2025." November 21, 2019. https://www.marketsandmarkets.com/Market-Reports/reality-applications-market-458.html.

Marr, Bernard. 2020. "The 5 Biggest Virtual and Augmented Reality Trends in 2020 Everyone Should Know About." *Forbes*, January 24, 2020. https://www.forbes.com/sites/bernardmarr/2020/01/24/the-5-biggest-virtual-and-augmented-reality-trends-in-2020-everyone-should-know-about/.

Marshmallow Laser Feast. 2016. *Treehugger: Wawona*. Mixed media. http://www.treehuggervr.com/.

Marshmallow Laser Feast. 2018. *We Live in an Ocean of Air*. Mixed media. https://www.marshmallowlaserfeast.com/experiences/ocean-of-air/.

Mascelli, Joseph V. (1965) 1998. *The Five C's of Cinematography: Motion Picture Filming Techniques*. Los Angeles: Silman-James Press.

Mashable. 2018. "This AR Experience Lets You Interrogate Hologram Passengers Like a Customs Officer." Interview by Mashable. YouTube video. August 31, 2018. https://youtu.be/1bkJtbEYdss.

Mateas, Michael, and Andrew Stern. 2003. "Façade: An Experiment in Building a Fully-Realized Interactive Drama." Paper presented at Game Developers Conference, San Jose, CA, March 4–8, 2003.

Matney, Lucas. 2018. "6D.Ai Is Building AR Tech That Crowdsources a 3D Mesh of the World." TechCrunch. February 20, 2018. http://social.techcrunch.com/2018/02/20/6d-ai-is-building-ar-tech-that-crowdsources-a-3d-mesh-of-the-world/.

Matsakis, Louise. 2019. "How the West Got China's Social Credit System Wrong." *WIRED*, July 29, 2019. https://www.wired.com/story/china-social-credit-score-system/.

Matsuda, Keiichi. 2016. *Hyper-Reality*. Film. http://hyper-reality.co/.

McCloud, Scott. 1993. *Understanding Comics: The Invisible Art*. New York: HarperCollins.

McCloud, Scott. 2000. *Reinventing Comics*. New York: HarperCollins.

McDougall, Walter A. 2013. *The Heavens and the Earth: A Political History of the Space Age*. Ann Arbor, MI: ACLS Humanities E-Book.

McIntosh, Jil. 2017. "How It Works: Ford's Virtual Reality Design Studios." Driving. June 14, 2017. https://driving.ca/auto-news/news/how-it-works-ford-design-studios.

McKeand, Kirk. 2020. "Half-Life: Alyx Review—VR's Killer App Is a Key Component in the Half-Life Story." VG247. March 23, 2020. https://www.vg247.com/2020/03/23 /half-life-alyx-review/.

McLuhan, Marshall. 1964. *Understanding Media: The Extensions of Man*. New York: New American Library Inc.

McMullan, Thomas. 2016. "Enter a Psychedelic Virtual Forest in Treehugger: Wawona." *Alphr*, December 14, 2016. https://www.alphr.com/art/1004973/enter-a-psychedelic -virtual-forest-in-treehugger-wawona.

Meehan, Michael, Sharif Razzaque, Brent Insko, Mary Whitton, and Frederick P. Brooks. 2005. "Review of Four Studies on the Use of Physiological Reaction as a Measure of Presence in Stressful Virtual Environments." *Applied Psychophysiology and Biofeedback* 30, no. 3: 239–258. https://doi.org/10.1007/s10484-005-6381-3.

Mer, Loren. 2012. "Virtual Reality Used to Train Soldiers in New Training Simulator." US Army. August 1, 2012. https://www.army.mil/article/84453/virtual_reality _used_to_train_soldiers_in_new_training_simulator.

Metacritic. 2020. "Half-Life: Alyx." https://www.metacritic.com/game/pc/half-life-alyx.

Microsoft. n.d. "Holoportation." Accessed December 30, 2018. https://www.microsoft .com/en-us/research/project/holoportation-3/.

Miesnieks, Matt. 2018. "Mapping the World with the AR Cloud." Presented at the Hardwired NYC, New York, NY, October 8, 2018. https://youtu.be/rlwJyg7uHGg.

Milgram, Paul, and Fumio Kishino. 1994. "A Taxonomy of Mixed Reality Visual Displays." *IEICE Transactions on Information and Systems* E77-D, no. 12: 1321–1329.

Milgram, Paul, Haruo Takemura, Akira Utsumi, and Fumio Kishino. 1995. "Augmented Reality: A Class of Displays on the Reality-Virtuality Continuum." In *Proceedings SPIE* 2351, Telemanipulator and Telepresence Technologies (December 21, 1995), edited by Hari Das, 282–292. Boston, MA: SPIE. https://doi.org/10.1117/12.197321.

Milk, Chris. 2015. "How Virtual Reality Can Create the Ultimate Empathy Machine." TED2015. March 2015. https://www.ted.com/talks/chris_milk_how_virtual_reality _can_create_the_ultimate_empathy_machine?language=en.

Mitchell, Robert. 1801. *Plans, and Views in Perspective. With Descriptions, of Buildings Erected in England and Scotland: And also an Essay, to Elucidate the Grecian, Roman and Gothic Architecture, Accompanied with Designs*. London: Oriental Press.

Monahan, Sean. 2020. "What Gay Men Can Teach Us about Surviving the Coronavirus | Sean Monahan." *The Guardian*, May 4, 2020. https://www.theguardian.com /commentisfree/2020/may/04/hiv-teach-us-to-survive-coronavirus.

Moore, Ben. 2020. "The Best VR Games for 2020." PCMAG. August 5, 2020. https://www.pcmag.com/picks/the-best-vr-games-for-2019.

Mori, Masahiro. 2012. "The Uncanny Valley: The Original Essay by Masahiro Mori—IEEE Spectrum." *IEEE Spectrum*, June 12, 2012. https://spectrum.ieee.org/automaton/robotics/humanoids/the-uncanny-valley.

Murray, Janet H. 1997. *Hamlet on the Holodeck: The Future of Narrative in Cyberspace.* Cambridge MA: MIT Press.

Nakamura, Lisa. 2020. "Feeling Good about Feeling Bad: Virtuous Virtual Reality and the Automation of Racial Empathy." *Journal of Visual Culture* 19, no. 1: 47–64. https://doi.org/10.1177/1470412920906259.

Narciso, David, Maximino Bessa, Miguel Melo, António Coelho, and José Vasconcelos-Raposo. 2019. "Immersive 360° Video User Experience: Impact of Different Variables in the Sense of Presence and Cybersickness." *Universal Access in the Information Society* 18, no. 1: 77–87. https://doi.org/10.1007/s10209-017-0581-5.

Naughton, John. 2018. "The Growth of Internet Porn Tells Us More about Ourselves than Technology." *Guardian*, December 30, 2018. https://www.theguardian.com/commentisfree/2018/dec/30/internet-porn-says-more-about-ourselves-than-technology.

New York Times. 2018a. "Augmented Reality: Explore InSight, NASA's Latest Mission to Mars." *New York Times*, April 30, 2018. https://www.nytimes.com/interactive/2018/05/01/science/mars-nasa-insight-ar-3d-ul.html.

New York Times. 2018b. "Step Inside the Thai Cave in Augmented Reality." *New York Times*, July 21, 2018. https://www.nytimes.com/interactive/2018/07/21/world/asia/thai-cave-rescue-ar-ul.html.

New York Times Magazine. 2016. "Man on Spire." *New York Times Magazine*, June 10, 2016. https://www.nytimes.com/2016/07/05/magazine/man-on-spire.html.

Nichols, Greg. 2018. "The Urgent Case for Open AR Cloud: Why We Need a Digital Copy of the Real World." ZDNET. June 7, 2018. https://www.zdnet.com/article/the-urgent-case-for-an-open-ar-cloud/.

Novarad. n.d. "OpenSight." Novarad. Accessed January 9, 2021. https://www.novarad.net/products/opensight/.

NYCT Subway (@NYCTSubway). 2016. "NYCT Subway on Twitter: "Hey #PokemonGO players, we know you gotta catch 'em all, but stay behind that yellow line when in the subway." Twitter, July 11, 2016, 2:00 p.m. https://twitter.com/NYCTSubway/status/752608535368769536/photo/1?utm_source=fb&utm_medium=fb&utm_campaign=gregjobe95&utm_content=752660047893782528.

Obar, Jonathan A., and Anne Oeldorf-Hirsch. 2020. "The Biggest Lie on the Internet: Ignoring the Privacy Policies and Terms of Service Policies of Social Networking Services." *Information, Communication & Society* 23, no. 1: 128–147. https://doi.org/10.1080/1369118X.2018.1486870.

Ochoa, Nick. 2015. "Jesse Schell's 40 Predictions for VR by 2025." UploadVR. November 10, 2015. https://uploadvr.com/jesse-schells-40-predictions-for-vr-by-2025/.

Oculus. 2019. "Welcome to Facebook Horizon." YouTube video. September 25, 2019. https://www.youtube.com/watch?v=Is8eXZco46Q.

O'Dea, S. 2020. "Smartphone Users Worldwide 2016–2021." Statista. December 10, 2020. https://www.statista.com/statistics/330695/number-of-smartphone-users-worldwide/.

OED Online. 2020. "Aesthetic, n. and adj." *OED Online*. December 2020. Oxford University Press. https://www.oed.com/view/Entry/3237#eid9579530.

Oettermann, Stephan. 1997. *The Panorama: History of a Mass Medium*. Trans. by Deborah Lucas Schneider. New York: Zone Books.

O'Hara, Kieron, and Wendy Hall. 2018. *Four Internets: The Geopolitics of Digital Governance*. CIGI paper no. 206. December 7, 2018. Waterloo: Centre for International Governance Innovation.

Omnicore. 2021a. "Facebook by the Numbers: Stats, Demographics & Fun Facts." *Omnicore* (blog). January 6, 2021. https://www.omnicoreagency.com/facebook-statistics/.

Omnicore. 2021b. "Instagram by the Numbers: Stats, Demographics & Fun Facts." *Omnicore* (blog). January 6, 2021. https://www.omnicoreagency.com/instagram-statistics/.

Ooi, Keng-Boon, Jun-Jie Hew, and Binshan Lin. 2018. "Unfolding the Privacy Paradox among Mobile Social Commerce Users: A Multi-Mediation Approach." *Behaviour & Information Technology* 37, no. 6: 575–595. https://doi.org/10.1080/0144929X.2018.1465997.

Osborne, Charlie. 2014. "How Not to Be a Glasshole? Google Explains." ZDNet. February 19, 2014. https://www.zdnet.com/article/how-not-to-be-a-glasshole-google-explains/.

Panetta, Kasey. 2018. "5 Trends Emerge in the Gartner Hype Cycle for Emerging Technologies, 2018." Smarter with Gartner. August 16, 2018. https://www.gartner.com/smarterwithgartner/5-trends-emerge-in-gartner-hype-cycle-for-emerging-technologies-2018/.

Panetta, Kasey. 2019. "5 Trends Appear on the Gartner Hype Cycle for Emerging Technologies, 2019." Smarter with Gartner. August 29, 2019. https://www.gartner .com/smarterwithgartner/5-trends-appear-on-the-gartner-hype-cycle-for-emerging -technologies-2019/.

Pangilinan, Erin, Steve Lukas, and Vasanth Mohan. 2019. *Creating Augmented and Virtual Realities: Theory & Practice for Next-Generation Spatial Computing*. Sebastopol, CA: O'Reilly Media.

Papacharissi, Zizi. 2010. *A Networked Self: Identity, Community, and Culture on Social Network Sites*. New York: Routledge.

Papagiannis, Helen. 2017. *Augmented Human: How Technology Is Shaping the New Reality*. Sebastopol, CA: O'Reilly Media.

Parisi, Tony. 2015. *Learning Virtual Reality: Developing Immersive Experiences and Applications for Desktop, Web, and Mobile*. Sebastopol, CA: O'Reilly Media.

Pauly, Olivier, Benoit Diotte, Pascal Fallavollita, Simon Weidert, Ekkehard Euler, and Nassir Navab. 2015. "Machine Learning-Based Augmented Reality for Improved Surgical Scene Understanding." *Computerized Medical Imaging and Graphics* 41 (April): 55–60.

Peterson, Andrea. 2016. "Holocaust Museum to Visitors: Please Stop Catching Pokémon Here." *Washington Post*, July 12, 2016. https://www.washingtonpost.com /news/the-switch/wp/2016/07/12/holocaust-museum-to-visitors-please-stop-catching -pokemon-here/.

Pew Research Center. 2019. "Mobile Fact Sheet." Pew Research Center: Internet and Technology. June 12, 2019. http://www.pewinternet.org/fact-sheet/mobile/.

Peyton, Lisa. 2018. "10 Ambitious Predictions for How VR/AR Will Shape Our World." VentureBeat. June 2, 2018. https://venturebeat.com/2018/06/02/10-ambitious-pre dictions-for-how-vr-ar-will-shape-our-world/.

Pixar. n.d. "Our Story; Pixar Animation Studios." Pixar Animation Studios. Accessed January 9, 2021. https://www.pixar.com/our-story-pixar.

"Pornography Is Booming during the Covid-19 Lockdowns." 2020. *The Economist*, May 10, 2020. https://www.economist.com/international/2020/05/10/pornography -is-booming-during-the-covid-19-lockdowns.

"Privacy Policy." n.d. PTC. Accessed July 25, 2020. https://www.ptc.com/en/docu ments/policies/privacy.

PR Newswire. 2018. "Global Virtual Reality Gaming Market 2018–2023: Market Expected to Grow at a CAGR of 26%." PR Newswire. Press release. September 10, 2018.

https://www.prnewswire.com/news-releases/global-virtual-reality-gaming-market
-2018-2023-market-expected-to-grow-at-a-cagr-of-26-300709527.html.

Ranger, Steve. 2020. "What Is the IoT? Everything You Need to Know about the
Internet of Things Right Now." ZDNet. Last updated February 3, 2020. https://www
.zdnet.com/article/what-is-the-internet-of-things-everything-you-need-to-know
-about-the-iot-right-now/.

Rebenitsch, Lisa, and Charles Owen. 2016. "Review on Cybersickness in Applications
and Visual Displays." *Virtual Reality* 20, no. 2: 101–125. https://doi.org/10.1007/s100
55-016-0285-9.

Rec Room Wiki. n.d. Rec Room Wiki home page. Fandom. Accessed June 7, 2019.
https://rec-room.fandom.com/wiki/Rec_Room_Wiki.

Reeves, Ben. 2020. "Valve On Why Half-Life: Alyx Needed To Be In VR." *Game In-
former.* March 3, 2020. https://www.gameinformer.com/interview/2020/03/03/valve
-on-why-half-life-alyx-needed-to-be-in-vr.

Reinhardt, Tilman. 2019. "Using Global Localization to Improve Navigation." *Google
AI Blog* (blog). Accessed December 28, 2019. http://ai.googleblog.com/2019/02
/using-global-localization-to-improve.html.

Reynolds, Emily. 2015. "The 'Uncanny Valley' Is Real, and Science Can Prove It."
WIRED, November 4, 2015. https://www.wired.co.uk/article/uncanny-valley-creepy
-robot.

Rheingold, Howard. 1991. *Virtual Reality*. New York: Summit Books.

Robert-Koch Institut. 2020. "Corona-Warn-App." June 16, 2020. https://www.corona
warn.app/en/.

Robertson, Adi. 2020. "Oculus Quest Games Are Getting Controller-Free Hand
Tracking This Month." *Verge*, May 18, 2020. https://www.theverge.com/2020/5/18
/21260554/oculus-quest-anniversary-hand-tracking-third-party-games-beat-saber
-tracks.

Robinson, Peter. 2016. "Remembering 'Pokemon Go,' the Craze That Swept July 2016."
VICE. September 7, 2016. https://www.vice.com/en_uk/article/xdmpgq/remember
ing-pokemon-go-the-craze-that-swept-july-2016.

Rogers, Ariel. 2013. *Cinematic Appeals: The Experience of New Movie Technologies*. New
York: Columbia University Press.

Rouse, Rebecca, Maria Engberg, Nassim JafariNaimi, and Jay David Bolter. 2015.
"MR^X: An Interdisciplinary Framework for Mixed Reality Experience Design and
Criticism." *Digital Creativity* 26, no. 3–4: 175–181. https://doi.org/10.1080/14626268
.2015.1100123.

Rousselle, Stefania, Veda Shastri, and Kaitlyn Mullin. 2016. "Lascaux Caves, Paleolithic and New Again." *New York Times*, December 18, 2016. https://www.nytimes.com/video/world/europe/100000004789226/lascaux-caves-paleolithic-and-new-again.html.

Rushton, Richard. 2011. *The Reality of Film: Theories of Filmic Reality*. Manchester, UK: Manchester University Press.

Ryzik, Melena. 2018. "Augmented Reality: David Bowie in Three Dimensions." *New York Times*, March 20, 2018. https://www.nytimes.com/interactive/2018/03/20/arts/design/bowie-costumes-ar-3d-ul.html.

Sabin, Dyani. 2017. "The Secret History of 'Pokemon GO,' as Told by Creator John Hanke." *Inverse*, February 28, 2017. https://www.inverse.com/article/28485-pokemon-go-secret-history-google-maps-ingress-john-hanke-updates.

Satariano, Adam. 2020. "How My Boss Monitors Me While I Work From Home." *New York Times*, May 6, 2020. https://www.nytimes.com/2020/05/06/technology/employee-monitoring-work-from-home-virus.html.

Sawh, Michael. 2020. "The Best AR Glasses and Smartglasses 2021: Snap, Vuzix and More." Wareable. October 22, 2020. https://www.wareable.com/ar/the-best-smart-glasses-google-glass-and-the-rest.

Schmalstieg, Dieter, and Tobias Höllerer. 2016. *Augmented Reality: Principles and Practice*. Boston: Addison-Wesley.

Schomer, Audrey. 2020. "Zuckerberg Outlines Facebook's Augmented Reality Ambitions." *Business Insider*, January 13, 2020. https://www.businessinsider.com/zuckerberg-outlines-facebook-augmented-reality-ambitions-2020-1.

Schwind, Valentin, Katrin Wolf, and Niels Henze. 2018. "Avoiding the Uncanny Valley in Virtual Character Design." *Interactions* 25, no. 5: 45–49. https://doi.org/10.1145/3236673.

Seppala, Timothy J. 2018. "The Auto Industry Is Head over Heels for VR." *Engadget*, January 26, 2018. https://www.engadget.com/2018/01/26/auto-industry-is-head-over-heels-for-vr/.

Shaw, Jeffrey. 1989. *The Legible City*. Interactive art installation. https://www.jeffreyshawcompendium.com/portfolio/legible-city/.

Sheffield Doc/Fest. 2018. "Artist Spotlight: Gabo Arora." Presented at the Sheffield Doc/Fest: Alternate Realities, Sheffield, UK, June 10, 2018. YouTube video. August 22, 2018. https://youtu.be/OEKMraIGOFI.

Sheridan, Thomas B. 1992. "Musings on Telepresence and Virtual Presence." *Presence: Teleoperators and Virtual Environments* 1, no. 1: 120–126. https://doi.org/10.1162/pres.1992.1.1.120.

Simonite, Tom. 2017. "Google's New Street View Cameras Will Help Algorithms Index The Real World." *WIRED*, September 5, 2017. https://www.wired.com/story /googles-new-street-view-cameras-will-help-algorithms-index-the-real-world/.

Singer, Natasha. 2020. "Virus-Tracing Apps Are Rife with Problems. Governments Are Rushing to Fix Them." *New York Times*, July 8, 2020. https://www.nytimes.com /2020/07/08/technology/virus-tracing-apps-privacy.html?referringSource=article Share.

Smithsonian Institution. 2014. *Skin & Bones*. iOS app. https://apps.apple.com/us/app /skin-bones/id929733243.

Snyder, Gabriel. 2017. "*The New York Times* Claws Its Way into the Future." *WIRED*, February 12, 2017. https://www.wired.com/2017/02/new-york-times-digital-journal ism/.

Solomon, Ben C., and Imraan Ismail. 2015. *The Displaced*. 360° film. New York Times Magazine. https://www.nytimes.com/2015/11/08/magazine/the-displaced-introduc tion.html.

Sorene, Paul. 2014. "Jaron Lanier's EyePhone: Head and Glove Virtual Reality in the 1980s." *Flashbak* (blog). November 24, 2014. https://flashbak.com/jaron-laniers-eye phone-head-and-glove-virtual-reality-in-the-1980s-26180/.

Stanney, Kay, Cali Fidopiastis, and Linda Foster. 2020. "Virtual Reality Is Sexist: But It Does Not Have to Be." *Frontiers in Robotics and AI*, January 31, 2020. https://doi .org/10.3389/frobt.2020.00004.

Statista. n.d. "Unit Shipments of Virtual Reality (VR) Devices Worldwide from 2017 to 2019 (in Millions), by Vendor." Statista. Accessed March 9, 2019. https://www .statista.com/statistics/671403/global-virtual-reality-device-shipments-by-vendor/.

Stein, Scott. 2019. "I Tried Facebook's Vision for the Social Future of VR, and It's Full of Question Marks." CNET. October 6, 2019. https://www.cnet.com/news/i-tried -facebooks-vision-for-the-social-future-of-vr-full-of-question-marks/.

Stein, Scott, and Ian Sherr. 2019. "Facebook's Zuckerberg Isn't Giving up on Oculus or Virtual Reality." CNET. September 25, 2019. https://www.cnet.com/features/face books-zuckerberg-isnt-giving-up-on-oculus-or-virtual-reality/.

Stephenson, Neal. 1992. *Snow Crash*. New York: Bantam Books.

Sterling, Bruce. 2010. "Augmented Reality: AR Uninvited at MOMA NYC." *WIRED*, October 6, 2010. https://www.wired.com/2010/10/augmented-reality-ar-uninvited -at-moma-nyc/.

Stoffregen, Thomas A, and L. James Smart. 1998. "Postural Instability Precedes Motion Sickness." *Brain Research Bulletin* 47, no. 5: 437–448. https://doi.org/10.1016 /S0361-9230(98)00102-6.

STRIVR. n.d. STRIVR home page. Accessed June 7, 2019. http://www.strivr.com/.

"Surveillance Camera Day: 20 June 2019." 2019. GOV.UK. May 15, 2019. https://www.gov.uk/government/publications/surveillance-camera-day-20-june-2019.

Sutherland, Ivan. (1965) 2009. "Augmented Reality: 'The Ultimate Display' by Ivan Sutherland, 1965." *WIRED*, September 20, 2009. https://www.wired.com/2009/09/augmented-reality-the-ultimate-display-by-ivan-sutherland-1965/.

Sutherland, Ivan E. 1968. "A Head-Mounted Three Dimensional Display." In *Proceedings of the December 9–11, 1968, Fall Joint Computer Conference, Part I on—AFIPS '68 (Fall, Part I)*, 757–764. San Francisco: ACM Press. https://doi.org/10.1145/1476589.1476686.

Taylor-Kroll, Charlie. 2016. "Pokemon Go Fan Becomes First Brit to 'Catch 'Em All.'" *Telegraph*, July 28, 2016. https://www.telegraph.co.uk/news/2016/07/28/pokemon-go-fan-becomes-first-brit-to-catch-em-all/.

TechFunnel. 2020. "The Complete Guide on Visual Positioning System." *TechFunnel* (blog). April 23, 2020. https://www.techfunnel.com/information-technology/visual-positioning-system/.

Tennent, Paul, Sarah Martindale, Steve Benford, Dimitrios Darzentas, Pat Brundell, and Mat Collishaw. 2020. "Thresholds: Embedding Virtual Reality in the Museum." *Journal on Computing and Cultural Heritage* 13, no. 2: 12:1–12:35. https://doi.org/10.1145/3369394.

Thiel, Tamiko. 2017. *Treasures of Seh Rem*. Augmented reality. May 27–November 30, 2017, Salem Maritime National Historic Site, Salem, MA. http://tamikothiel.com/tosr/index.html.

Thiel, Tamiko, with /p. 2019. *Strange Growth*. Augmented reality. http://tamikothiel.com/AR/strange-growth.html.

Tuan, Yi-Fu. (1977) 2001. *Space and Place: The Perspective of Experience*. Minneapolis: University of Minnesota Press.

Turkle, Sherry. 1995. *Life on the Screen: Identity in the Age of the Internet*. New York: Simon & Schuster.

US Army. n.d. "Synthetic Training Environment (STE)." US Army. Accessed June 7, 2019. https://asc.army.mil/web/portfolio-item/synthetic-training-environment-ste/.

USC Institute for Creative Technologies. 2005–present. "Bravemind: Virtual Reality Exposure Therapy." https://ict.usc.edu/prototypes/pts/.

Valentino-DeVries, Jennifer, Natasha Singer, Michael H. Keller, and Aaron Krolik. 2018. "Your Apps Know Where You Were Last Night, and They're Not Keeping It

Secret." *New York Times*, December 10, 2018. https://www.nytimes.com/interac tive/2018/12/10/business/location-data-privacy-apps.html.

Verykokou, Styliani, Charalabos Ioannidis, and Georgia Kontogianni. 2014. "3D Visualization via Augmented Reality: The Case of the Middle Stoa in the Ancient Agora of Athens." In *Digital Heritage: Progress in Cultural Heritage: Documentation: Preservation, and Protection*, 279–289. New York: Springer.

Viator. 2017. "Best TV & Film Tours around the World." *HuffPost* (blog). December 6, 2017. https://www.huffpost.com/entry/best-tv-film-tours_b_3386410.

Volkswagen. n.d. "What Does a VR Developer Actually Do, Mr. Kuri?" Accessed January 9, 2021. https://www.volkswagenag.com/en/news/stories/2018/05/what-does -a-vr-developer-actually-do-mr-kuri.html#.

Voros, Joseph. 2003. "A Generic Foresight Process Framework." *Foresight* 5, no. 3: 10–21. https://doi.org/10.1108/14636680310698379.

Wagner, Kurt. 2019. "Facebook Almost Missed the Mobile Revolution. It Can't Afford to Miss the Next Big Thing." Vox. April 29, 2019. https://www.vox.com/2019 /4/29/18511534/facebook-mobile-phone-f8.

Weiser, Mark. 1991. "The Computer for the 21st Century." *Scientific American*, September 1991. https://www.lri.fr/~mbl/Stanford/CS477/papers/Weiser-SciAm.pdf.

Weiser, Mark, and John Seely Brown. 1996. "The Coming Age of Calm Technology." October 5, 1996. https://calmtech.com/papers/coming-age-calm-technology.html.

White, Sean. 2018. "Enabling Social Experiences Using Mixed Reality and the Open Web." *The Mozilla Blog: Dispatched from the Internet Frontier.* April 26, 2018. https:// blog.mozilla.org/blog/2018/04/26/enabling-social-experiences-using-mixed-reality -and-the-open-web/.

Wikipedia contributors. 2020a. "Google Earth." Wikipedia. Version last updated July 21, 2020. https://en.wikipedia.org/w/index.php?title=Google_Earth&oldid=968784 015.

Wikipedia contributors. 2020b. "Google Maps." Wikipedia. Version last updated July 21, 2020. https://en.wikipedia.org/w/index.php?title=Google_Maps&oldid=968842 334.

Wikipedia contributors. 2020c. "Google Street View." Wikipedia. Version last updated July 15, 2020. https://en.wikipedia.org/w/index.php?title=Google_Street_View&old id=967756690.

Wikipedia contributors. 2020d. "History of Television in Germany." Wikipedia. Version last updated June 16, 2020. https://en.wikipedia.org/w/index.php?title=History _of_television_in_Germany&oldid=962906935.

Wikipedia contributors. 2020e. "John Logie Baird." Wikipedia. Version last updated July 4, 2020. https://en.wikipedia.org/w/index.php?title=John_Logie_Baird&oldid =966048213.

Wikipedia contributors. 2020f. "List of Video Game Genres." Wikipedia. Last updated December 6, 2020. https://en.wikipedia.org/wiki/List_of_video_game_genres.

Wikipedia contributors. 2020g. "Oculus Rift." Wikipedia. Version last updated June 21, 2020. https://en.wikipedia.org/w/index.php?title=Oculus_Rift&oldid=963755597.

Wikipedia contributors. 2020h. "Pokémon Go." Wikipedia. Version last updated May 11, 2020. https://en.wikipedia.org/w/index.php?title=Pok%C3%A9mon_Go&oldid =956111887.

Wikipedia contributors. 2020i. "HTTP Cookie." *Wikipedia*. Version last updated May 14, 2020. https://en.wikipedia.org/w/index.php?title=HTTP_cookie&oldid=956554 418.

Williams, Raymond. 2004. *Television: Technology and Cultural Form*. 3rd ed. New York: Routledge.

Xiao, Eva. 2017. "WeChat Is Quietly Developing Its Own AR Platform." Tech in Asia. September 15, 2017. https://www.techinasia.com/wechat-first-look-ar-platform.

Yamagata Corp. 2017. "Hologarage: Microsoft Hololens Car Maintenance Demo." YouTube video. March 14, 2017. https://www.youtube.com/watch?v=5HV3fcTvZk0.

Yeo, Elizabeth, Brian Chau, Bradley Chi, David E. Ruckle, and Phillip Ta. 2019. "Virtual Reality Neurorehabilitation for Mobility in Spinal Cord Injury: A Structured Review." *Innovations in Clinical Neuroscience* 16, no. 1–2: 13–20.

Young, Chris. 2019. "Virtual Reality Can Help Relieve Severe Pain in Patients, Study Finds." Interesting Engineering. August 15, 2019. https://interestingengineering.com /virtual-reality-can-help-relieve-severe-pain-in-patients-study-finds.

Youngblood, Gene. 1970. *Expanded Cinema*. New York: E. P. Dutton,

Zephoria. 2020. "The Top 20 Valuable Facebook Statistics—Updated October 2020." Zephoria Digital Marketing. October 2020. https://zephoria.com/top-15-valuable -facebook-statistics/.

Zone, Ray. 2007. *Stereoscopic Cinema and the Origins of 3-D Film, 1838–1952*. Lexington: University Press of Kentucky.

Zone, Ray. 2012. *3-D Revolution: The History of Modern Stereoscopic Cinema*. Lexington: University Press of Kentucky.

Zuboff, Shoshana. 2019. *The Age of Surveillance Capitalism: The Fight for a Human Future at the New Frontier of Power*. New York: PublicAffairs.

Index